Guide to Inshore Marine Life

David Erwin

Bernard Picton

Reprinted 1990

Phototypeset in Plantin by
Datapage International, Dublin

Printed and bound in Japan by
Dai Nippon Printing Co., Tokyo

Production and design by
Jane Stark, Connemara Graphics, Ireland

ISBN 0 907151 345

IMMEL Publishing
Ely House
37 Dover Street
London W1X 3RB

CONTENTS

FOREWORD
by Bob Earll,

Anyone who has sat beside a rock pool, been snorkelling or diving in the sea, will be fascinated by the wealth of marine life they have seen. With their appetites whetted they will go to the book shelf to find a marine life guide to explain what they have seen. At that point they will be disappointed to find that the illustrations often bear little or no resemblance to what they have seen and that the scientific jargon used confuses rather than enlightens. The Marine Conservation Society Guide to the Inshore Marine Life aims to change that.

The authors have with their considerable talents sought to break new ground to assist visitors to the sea shore. David Erwin and I first collaborated on a project to enable divers to record the marine life they saw whilst underwater. Many of the ideas he pioneered then are developed in this guide. In the organisation of this guide he has deliberately set out to break with convention in a way which builds on peoples experience, and which assists them to get quickly to the photographs which are most likely to be relevant to their interest. The photographs are superb and have been taken by Bernard Picton, one of the leading photographers of marine life in this country. The photographs of living species show just how spectacular these organisms are when they are alive. They also show their natural habitats, a feature which often considerably aids indentification.

The ideas behind this book have been forming for a number of years, and considerable effort has gone into the selection of the species which represent some of the most conspicuous and common species to be found in European waters. Much of the information used in the guide has come from the work of volunteer divers who have taken part in many of the Marine Conservation Society projects and the guide is a fitting tribute to their considerable activities and enthusiasm.

The efforts of the Marine Conservation Society and its members are aimed at helping to explain the wonders of the marine world. We hope that this guide will help you to enjoy the marine life you see, and that it will encourage you to share these experiences with others to further the cause of marine conversation.

Dr. Bob Earll
General Secretary of the Marine Conservation Society.

PREFACE

Many academic publications and monographs are available for the definitive identification of the Marine Life of the North East Atlantic. For the majority of people, however most of these writings are either inaccessible or incomprehensible. They provide a way for the experienced professional marine biologist to identify material but they do not give any help to the beginner or the enthusiast without the benefit of formal training. They deal with the subject at a level of complexity far beyond the needs of the average reader, often acting as a barrier to developing interest. The animals and plants described seem to be remote and obscure. Yet in the marine environment the organisms are far from obscure. We encounter them "face to face" every time we dive in the sea or look into a rock pool. We feel a need to be able to put a name on what we are seeing and perhaps learn a little about it.

This book uses photographs taken underwater in an attempt to make that possible. It is not intended as a comprehensive identification manual but rather as an introduction to the wealth of marine life found in the shallow water around our coasts. The 200 species illustrated are not the only animals or plants you can expect to find but they are some of the more easily recognised ones. Anywhere on the coast, on every dive and in any rock pool, you will always be able to find something illustrated within these covers.

LIFE IN THE SEA
A Conglomeration of Shapes and Colours?

When we look into the marine environment for the first time we quickly become confused by an apparent conglomeration of shapes and colours.

It is rather like when we go to a hospital to visit a sick friend and on the way in we become part of a seething and amorphous mass of people—that we know nothing about—or do we? From previous visits or from other experiences we have an idea of the sort of people who work in hospitals. We see large numbers of people dressed in different ways. Without really thinking we spot white coats and rightly conclude that we are seeing doctors. We recognise nurses and can even start to say what type or rank of nurse we are looking at, depending on details of the uniform worn. We recognise ancilliary staff and correctly separate different jobs within that general heading. All of this we do in a background of a great many more people who remain as part of an undifferentiated and undistinguishable "crowd". We have used "clues" to spot and identify the groups.

In the sea for the first time we do not usually have the advantage of previous experience. On return to the surface we say something like: "There was so much—fantastic colours, terrific shapes—but I do not know what any of it was". In fact we have looked at everything but we have seen nothing. If pressed we will remember, "Oh yes, there was a fish that swam by"—and that is important.

Why did we see and remember the fish? How did we separate it out from all the other things? Amongst other things it was moving, we recognised it as "a fish" and in the sea we expected to see fish.

Through this book you will soon be able to recognise some of the other major groups of marine animals and plants. In addition you should quickly learn what, in general terms, you might expect to find in different marine habitats and with practice you should be able to put a name on many of the organisms you see around the coast.

HOW TO USE THIS BOOK

The book is intended as a practical aid to the answering of two questions:

WHAT IS IT?

(What group it is in and what is its name?)

and

WHERE DOES IT LIVE?

(What kind of place, what sort of bottom and at what depth?)

What is it?

The next four pages show examples of ten of the main animal and plant groups present in shallow water around the coast. Six are dealt with individually and four are treated together as a loose assortment of animals which are perhaps less obvious.
Each is accompanied by a short description of the GROUP or groups involved.
This does not mean that these are the only GROUPS to be found. They are however the ones you are most likely to see. They are not laid out in what biologists would call 'systematic order' and no status or relationship is implied.
Each photograph is set under a strip of a colour which has been allocated throughout the book to members of that group and appears at the top and bottom of relevant pages. You can use it to quickly find the section you want or to refer back to this section.
For example—FISH are given BLUE as their colour. If you want to go to the fish section—flick through until you find the 'blue pages'. These all show fish. In the same way the MOLLUSCS are on the YELLOW pages and so on.

Centrolabrus exoletus, the Rock cook, a common fish of kelp forest and shallow rock.

Fish have:

- An obvious head with eyes and a mouth.
- 'Fins' at the tail and usually on the back.
- An opening or series of openings on each side for water to pass over the gills.
- Usually a covering of 'scales' or tiny teeth.

Several different echinoderms on an exposed rock reef.

Echinoderms have:

- A body based on a pattern of '5' (with a few exceptions!).
- Bright coloured skin which is often hard to the touch.
- A large number of writhing 'tube-feet', part of a unique hydraulic movement system (sometimes too small to see).
- Hard spines in rows or all over the body.

Starfish, Urchins, Brittlestars, Featherstars and Sea-cucumbers

A row of *Cancer pagurus*, the edible crab on a rock ledge.

Crustaceans have:
- A hard outer horny "shell".
- Legs made up of several parts which joint together.
- Obvious eyes, sometimes on stalks.
- Obvious body segments (look under a crab!).

Urticina felina—a common sea anemone on shallow rock. The very similar but larger *Urticina eques* is a close relative found on mud and gravel.

Cnidarians (pronounced nidarians) have:
- A 'flower-like' appearance.
- Tentacles which trap food.
- A wide range of bright colours.
- Stinging cells!

Snails, Sea-Slugs, Clams and Octopus

A close up photograph of the 'head' of *Aporrhais pespelecani*, the Pelican's foot shell (see page 69).

Molluscs have:
- Chalky shells (except octopus and sea slugs).
- A wide range of colour and pattern of shell.
- A fleshy body—often yellow.
- *NO* obvious body segments.

Shallow rock on St Kilda with a wide variety of algae growing on it.

Algae have:
- Green, brown or red coloration.
- A covering of mucus and a slippery 'feel'.
- A means of attachment to the bottom (or something else!).
- Often a flat or feathery appearance.

Kelps and Seaweeds

This is a completely artificial group of unrelated animals bunched together purely for convenience. Even 'worms' is only a descriptive term rather than a biological one and includes several scientific animal groupings.

Haliclona viscosa—a Sponge
see pages 85 to 91

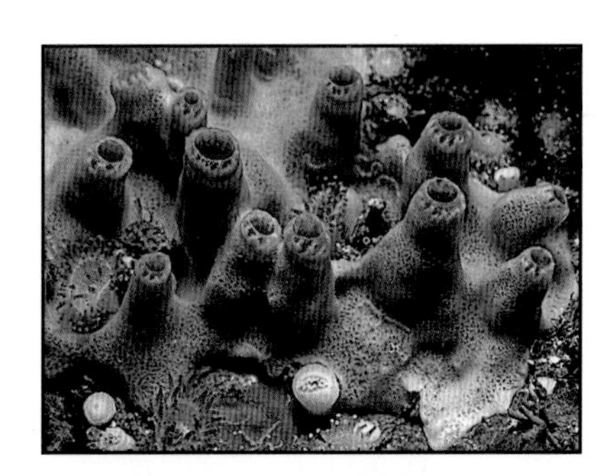

Phyllodoce paretti—a worm
see pages 92 to 96

Securiflustra securifrons
A Bryozoan or Sea Mat
see pages 97 to 99

Ciona intestinalis
a Tunicate or Sea Squirt
see pages 100 to 104

Don't be afraid of latin names!

Most animals and plants found in the sea do not have a 'common name' and many of those which do have one often rejoice in a series of names, each peculiar to a particular place or region. For example, which of the following fish do you know?

Saithe	Coalfish	Blockan	Glashen
Gilpen	Greylord	Blackfish	Coley

In fact they are all the same thing. They are local names for one species of fish *Pollachius virens* (page 20)—all from within a short stretch of coastline.

Wherever you go however, the latin (or scientific) name stays the same. In Belfast, Berlin or Bangkok *Pollachius virens* **always** refers to the same fish. The rules of Biology totally prohibit the same name being given to anything else.

How do latin names work?

Latin or scientific names all have two parts.

The **genus** always comes first and always has a capital or upper case first letter. It is the equivalent of our Surname.

The **species** always comes second and always has a small or lower case first letter. It is the equivalent of our Christian or first name.

Together these two names, or 'binomial' form the full name of an organism, e.g.

Cancer pagurus—The edible crab (page 47).

Cancer is the Genus *pagurus* is the species.

In a written text scientific names are usually separated out as above by the use of italics so that no confusion can exist.

Closely related **species** may have the same **generic** name but different **specific** names—like *Pollachius virens*, (page 20) the coalfish and *Pollachius pollachius*, (page 19) the pollack.

Closely related Genera can be grouped together into Families, Families into Orders, Orders into Classes and Classes into Phyla. However none of these classifications or groupings are required to name an animal or plant. They are all merely conveniences for us to be able to group similar things together.

Don't worry about how to pronounce a latin name.

Have a try!

Say it as it looks.

You have a good chance of being right.

Where does it live?

1. What sort of place?

These pages show examples of a range of underwater sites from exposed or current swept sites which are subjected to a lot of water movement to sheltered or still sites with very little water movement. They are intended to convey the general impression or 'feel' of the sites.

To the right of each photograph is a spectral scale coloured from red to blue. Red represents **high energy** where there is a lot of wave action or current. Blue represents **low energy** where there is little wave action or current.

Each of the identification photographs of animals and plants later is accompanied by this scale with the parts of the range where they are usually found marked. Reference back to this section will give an approximate idea of the type of place it can normally be found.

a) High energy site

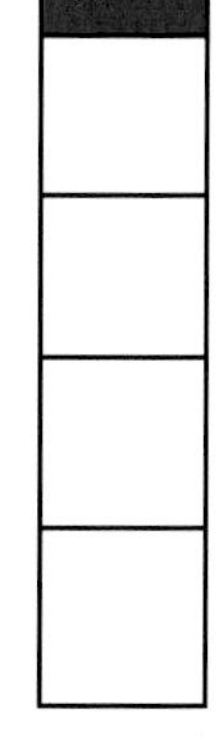

Bedrock or large boulders, with no silt. Species which abound include the anemones *Corynactis viridis* (page 56) and *Metridium senile* (page 58) together with a variety of encrusting sponges. There is nothing which projects very far into the water.

b) Moderately high energy site

Bedrock, boulders and cobbles, sometimes with coarse sediment under and between. Dense populations of sponges, sea squirts and hydroids. The soft coral *Alcyonium digitatum* (page 52) and the anemone *Actinothoe sphyrodeta* (page 59) are often abundant.

c) Mid-energy site

Cobbles and occasional boulders on coarse sand or muddy sand. Sparse hydroids, sponges and sea squirts on rocks with crabs and squat lobsters common under rocks. Brittle stars *Ophiothrix fragilis* (page 33) or *Ophiocomina nigra* (page 34) often abundant.

d) Moderately low energy site

Largely sand plains. May have wave ripples or current formed sand waves. Little on surface except around an occasional stone or boulder. Flatfish and shrimps may be very common and the large anemone *Urticina eques* is often present.

e) Low energy site

Mud or fine sand plains. Often a lot of evidence for activity under the surface in the form of casts and burrows. The crab *Liocarcinus depurator* (p 46) is often present and sea pens (page 54) regularly are found standing free of the sediment.

2. On, in or around what sort of bottom?

A series of icons or symbols runs along the bottom of each photograph, depicting the type of bottom or habitat in, on or around which the species is often found.

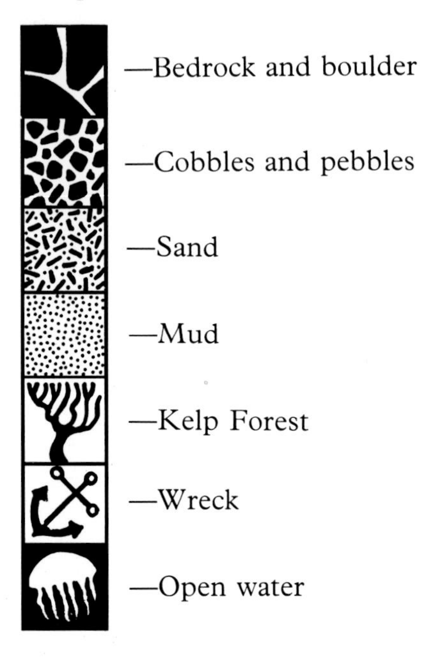

3. How deep?

To the left of each identification photograph is a depth scale showing the range of depth in which you would normally expect to find the animal or plant.

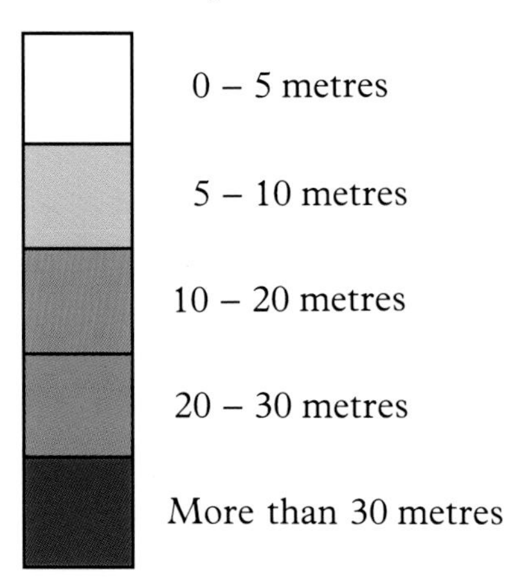

If we add an indication of size of the organism or colony;

○ —50P piece
○ —hand
○ —suitcase

and put it all together we arrive at the following:

FISH

Lipophrys pholis—Shanny

From it we can see that:

- This is a Fish called *Lipophrys pholis*—the Shanny, which
- is about the size of a hand which
- lives in shallow water, down to about 10m in depth is found
- in rocky areas and amongst kelp where
- there is a moderate amount of wave action or current.

We also have in each case, in note form, key identification points, where relevant, or simply an interesting fact or two about the animal or plant. For example:

Lipophrys pholis—Shanny

Abundant amongst weed in rock pools and shallow water on all coasts. Feed on shrimps and small crabs. Young shannies eat the extended legs of feeding barnacles.

Conger conger—Conger eel

Hide under rocks, in crevices and in wrecks during daylight hours. Feed voraciously on anything that moves including quite large fish.

Sygnathus acus—Greater pipefish

Found amongst weeds and eel grass over sand and mud bottoms. Looks like a straightened sea-horse and swims with its body vertical in water.

Trisopterus luscus—Pouting or Bib

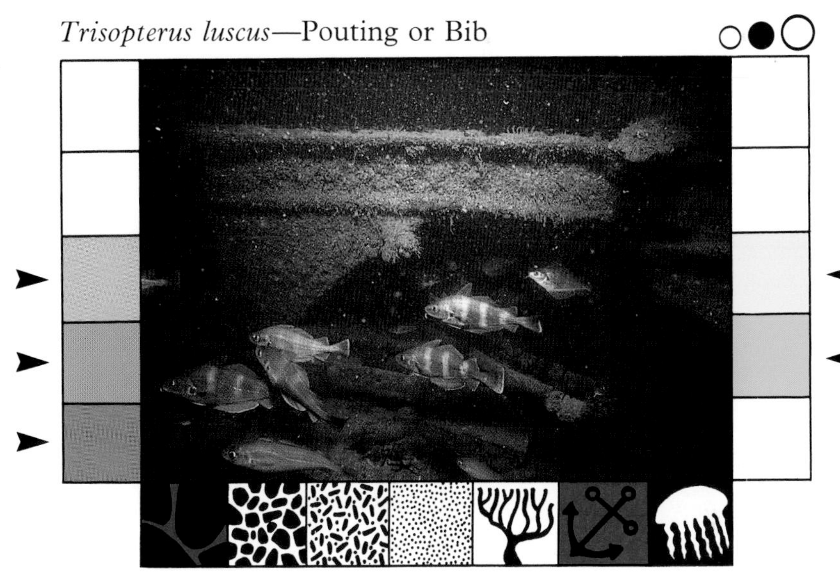

Common shallow water fish. Sometimes found in large shoals. Often found in and around wrecks or swimming near rock outcrops in sandy areas.

Pollachius pollachius—Pollack or Lythe

Common shoaling fish associated with wrecks, rocks and reefs. Lower jaw projects beyond the upper jaw. Line along each side (lateral line) is dark and curves up at front end.

Pollachius virens—Saithe or Coley

Very common smaller relative of Pollack. Shoals over kelp, in rocky areas and around wrecks. Jaws of equal length. 'Lateral line' light coloured and straight.

Molva molva—Ling

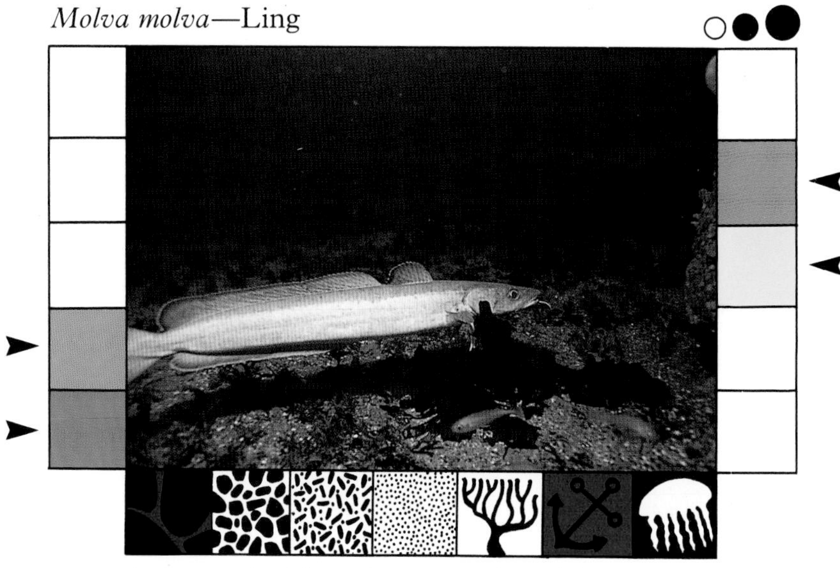

Associated with rocks or wrecks subject to a lot of water movement. Voracious predator of fish. Single beard (barbel) is held straight forward, sensing the water ahead.

Labrus bergylta—Ballan wrasse

All Ballan wrasse are females for their first eight years. A few then change internally into fully functional males. Large Ballans are almost certainly males.

Labrus mixtus—Cuckoo wrasse

Females are reddish with three dark spots near the tail. Males have a bright blue head and streaky blue and orange sides. Head has a white patch when spawning.

Ctenolabrus rupestris—Goldsinny wrasse

Very common singly or in small shoals in sandy or rocky areas. Characteristic black spot on top of tail at base of fin. Has large obvious scales.

Callionymus lyra—Dragonet

Spend a lot of time lying still on sand or mud bottoms and can be difficult to spot. Males have a spectacular dorsal fin almost as long as the body.

Chirolophis ascanii—Yarrell's blenny

Lives in cracks and crevices in rock at all depths. Characteristic large fringed tentacles, one above each eye. Two more small tentacles behind head on fin. In south, the Tompot blenny is more common.

Pholis gunnellus—Butterfish

Very common under rocks on shores and in rock pools. Also found deeper amongst cobbles and pebbles where there is a lot of water movement.

Thorogobius ephippiatus—Leopard-spotted goby

Was thought to be very rare before diving became popular. Now known to be widespread and quite common in holes and crevices at the base of steep rock faces.

Gobiusculus flavescens—Two-spotted goby

Common in small shoals amongst weed or eel grass. Distinctive pale-edged dark spot at base of tail. Males only have another spot on side behind fin.

Taurulus bubalis—Sea scorpion

Lives on lower shores, rock pools and in shallow water amongst rocks where weed is present. Feeds on fish, shrimps and crabs.

Lophius piscatorius—Angler Fish

Lie absolutely still on sand and gravel bottoms. Mouth quickly opens to ensnare fish. Can grow to extremely large sizes. Do not put your hand in its mouth!

Pleuronectes platessa—Plaice

Most commonly found on sand and gravel but also occasionally on mud. Often partially cover themselves with sediment to avoid detection. Will eat almost anything.

Zeugopterus punctatus—Topnot

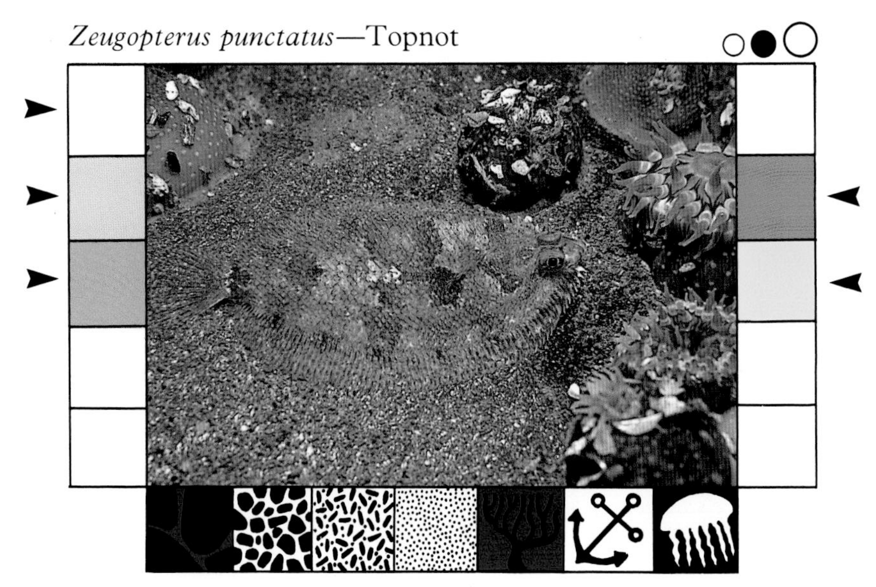

Small flatfish found in shallow rocky areas and amongst weed. Seem to like a lot of water movement. Can 'cling' to rocks, even upside down on overhangs.

Scyliorhinus canicula—Lesser spotted dogfish

Usually found on or close to the bottom. Feeds on molluscs, crabs and small fish. Egg capsule, mermaid's purse, is often found attached to weed in shallow water.

Raja clavata—Thornback ray

Found on sand or mud. Feeds on flatfish, shrimps, prawns and crabs. Sharp spines on upper surface of body especially in 3 rows on tail.

Antedon bifida—Featherstar

Found, sometimes in very large numbers, on rock, weed and sessile animals. Ten very mobile flexible arms held up into the water. Colour is variable from purple to orange.

Astropecten irregularis—Sand star

Common and widespread on, or just under, the surface of sand or sandy mud. Characteristic stiff, flat appearance with a fringe of light plates and spines.

Luidia ciliaris—Seven-armed starfish

Seven orange arms fringed with prominent white spines. Lives on sand, gravel or sand scoured rock over which it can move relatively quickly.

Porania pulvillus—Red cushion-star

Lives on wave affected or current swept rocks where the soft coral *Alcyonium digitatum* (page 52) is present. Sometimes found on wrecks.

Asterina gibbosa—Cushion star or Starlet

Small starfish commonly found under or pressed close to the surface of weed-covered intertidal rocks and boulders. Also subtidally in sheltered areas.

Anseropoda placenta—Goosefoot starfish

Lives in or on deeper muddy sand or muddy gravel. Found all around the British Isles but not often seen by divers. Does it bury itself in the bottom?

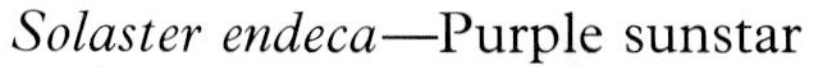

Solaster endeca—Purple sunstar

On current swept rock or in areas with silt covered rock and boulders surrounded by muddy gravel. Smooth surface which is hard to the touch. Usually 9 or 10 arms.

Crossaster papposus—Common sunstar

Common in current swept sediment sites or wave exposed rock sites. Sometimes small specimens are found on the shore. Eat other starfish, sometimes another Common sunstar.

Henricia oculata—Bloody henry

Common in the most wave exposed or tide swept sites in the south. However difficult to be sure of identification as in the north the almost identical *Henricia sanguinolenta* is common.

Asterias rubens—Common starfish

Widespread and common. Feed largely on molluscs and are particularly common on mussel beds where they can grow very large. Occasionally in large numbers.

Marthasterias glacialis—Spiny starfish

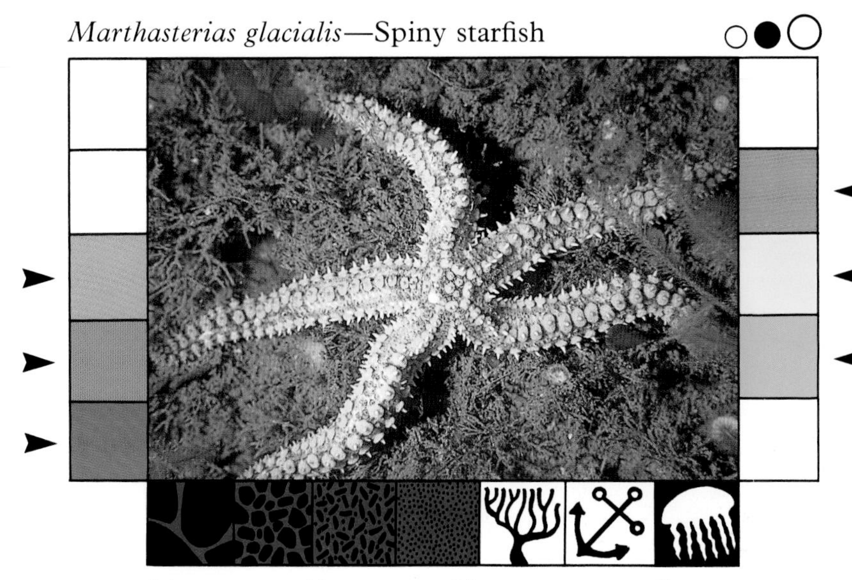

Live on a wide range of bottom types. Sometimes can be very large (up to 75cms in diameter), especially on mud. Possibly have a southern distribution?

Ophiothrix fragilis—Common brittlestar

Found beneath boulders but can also exist in vast dense beds of many millions of individuals. Very variable in colour but arms always banded, light and dark.

Ophiocomina nigra—Black serpent-star

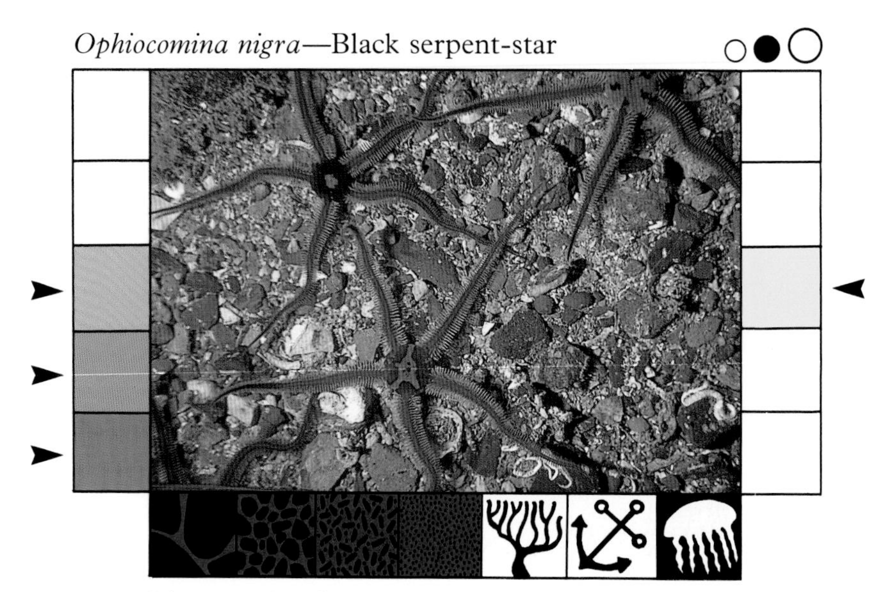

Live out in the open on rock or gravel areas with some water movement. Sometimes found in beds on sediment but only moderately dense compared to *Ophiothrix fragilis*.

Ophiopholis aculeata—Crevice brittlestar

Live in crevices, cracks and holes, natural or bored in cobbles boulders or bedrock. Often just legs can be seen. Red or bluish banded pattern.

Ophiura texturata—Large sand brittlestar

Writhe across the surface of sand or muddy sand. Can move fast with a swimming action when disturbed. Occasionally bury themselves just below the sand surface. Up to 10cm diameter.

Psammechinus miliaris—Shore urchin

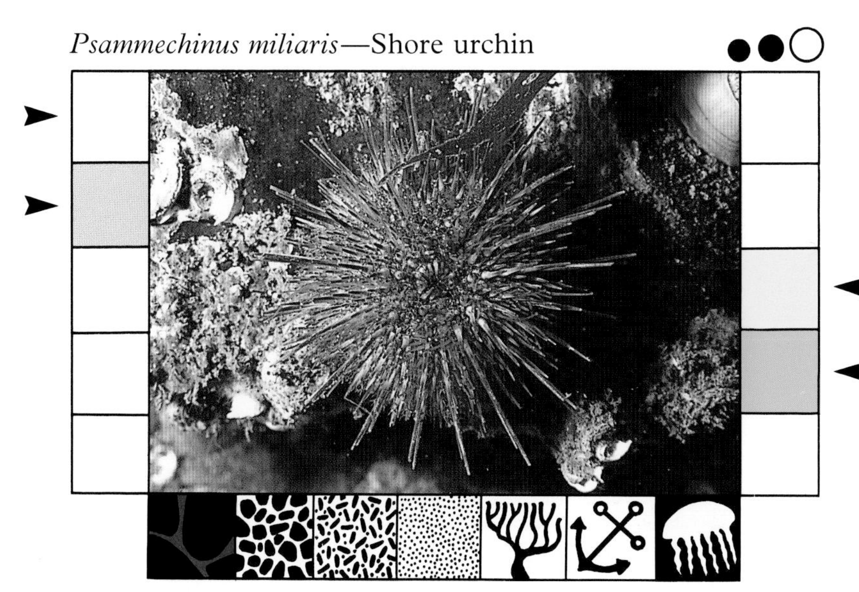

Found in rock pools and on the lower edge of rocky shores with weed cover. Subtidally in sea loughs or other sheltered sites. Usually about 5cm diameter.

Echinus esculentus—Common sea urchin

Probably the most important grazer of subtidal rock surfaces—vital to maintenance of diversity in an area. Much sought after for tourist and curio trade.

Paracentrotus lividus—Black urchin

West coast species. Lives in hollows or 'crypts' in limestone. Occasionally dense beds in maerl. Now absent from many previously rich areas due to stripping for French food markets.

Strongylocentrotus droebachiensis—Northern urchin

On rocky shores and subtidally in sheltered sites. Also often recorded where currents are fairly strong. North east Scotland, Shetlands, Faroes and Scandinavia.

Spatangus purpureus—Purple heart-urchin

Widespread just below the surface of gravel and coarse shell sand. Tell-tale furrows show where an urchin has moved along. Large, up to 12cm in length.

Echinocardium cordatum—Common heart-urchin

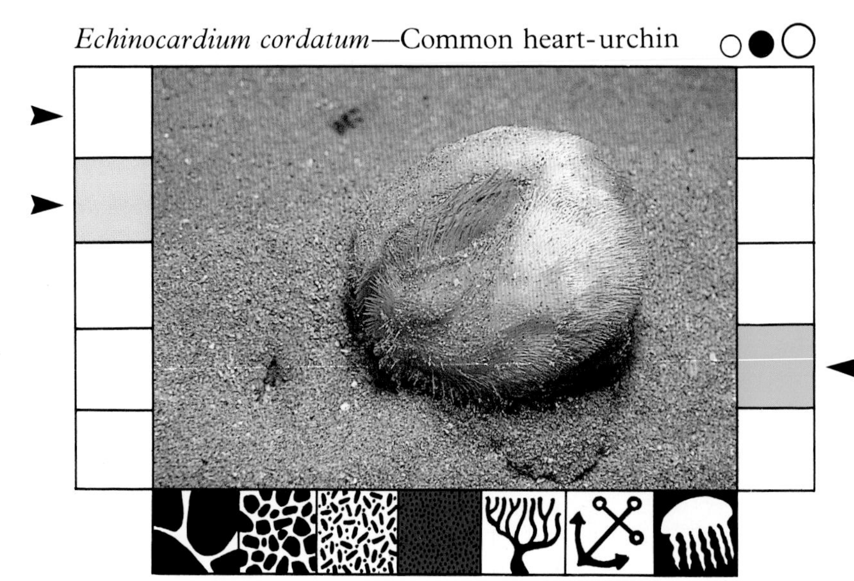

Live about 12cm below the surface of clean sand. Detected by a conical depression in the sand above the animal where detritus collects and is drawn down for food.

Holothuria forskali—Cotton spinner

Common on southern and some north western rocky sites where there is a lot of wave exposure. If disturbed will readily throw out sticky white threads.

Aslia lefevrei—Brown sea-cucumber

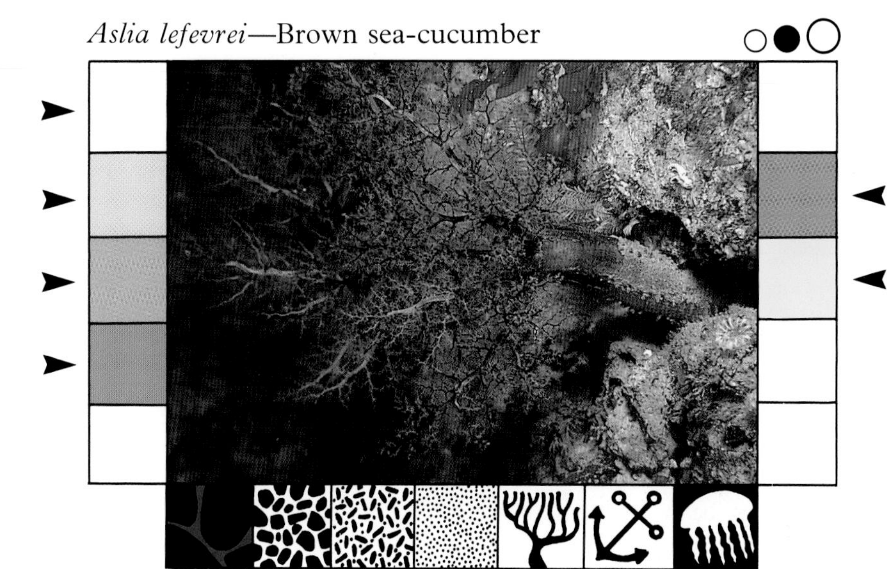

Widespread in rock crevices down to 30m. Typically brownish body in crevice with darker feeding tentacles spread out in water. Pure white body is *Pawsonia saxicola*.

Thyone roscovita—Sea-cucumber

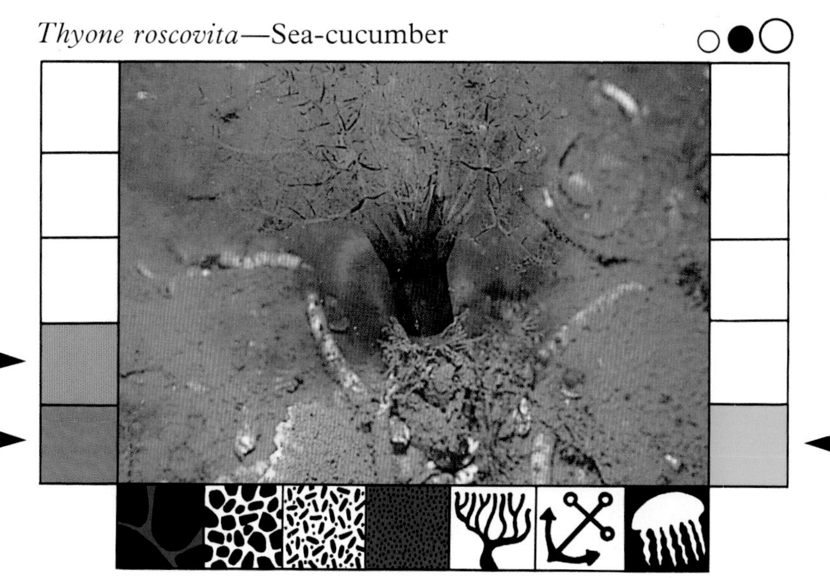

Widespread. Dark brown body buried just below the surface of mud or muddy sand. Delicate dark aborescent tentacles with light patches, held up in water. No light patches—probably *Thyone fusus*.

Neopentadactyla mixta—Gravel sea-cucumber

Found, sometimes in large numbers, in gravel or maerl subject to strong currents. Mystery as to why all sometimes disappear below the surface at the same time!

Cucumaria frondosa—Pudding

In shallow water from north east Scotland to Scandinavia. Sticks its very dark leathery body to rocks with bushy tentacles in water. Very large—up to 50cm in length.

Nephrops norvegicus—Dublin Bay prawn or Scampi

Often found sitting at one entrance to its burrow in the finest of muds in deep water. Can occur in shallow water in very sheltered areas. Sometimes in large numbers.

Homarus gammarus—Lobster

Normally in subtidal rocky areas subject to a lot of water movement but sometimes in rock pools or sandy areas. Nocturnal in shallow water, usually hiding during the day.

Palinurus elephas—Crawfish

Most common on southern and western exposed rocky coasts. Prefers boulders or bedrock and can be numerous on vertical cliffs. Seasonal—go deeper offshore in winter.

Munida rugosa—Long-clawed squat lobster

Very common in sea loughs and on deep sandy bottoms on open coast. Nocturnal in shallow water, during the day hiding under rocks with just the claws sticking out.

Galathea strigosa—Spiny squat-lobster

Common in crevices and under rocks on lower shores and subtidal rocky areas with a lot of water movement. Can be extremely aggressive when caught!

Pagurus bernhardus—Common hermit crab

Found almost anywhere, on any bottom, at any depth. Need to find increasingly larger shells as they grow. When small in a wide variety but when large usually in whelk shells.

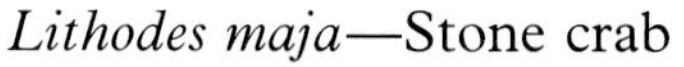
Lithodes maja—Stone crab

A northern species. At its southern limit in deep water exposed sites on the west coast of Scotland and rarely, Ireland. Also occasionally in sea loughs. Characteristic large spines all over body.

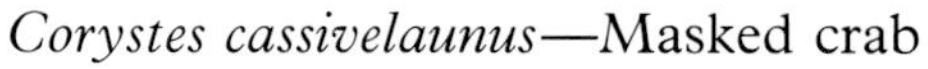
Corystes cassivelaunus—Masked crab

Common and widespread in sand at all depths. During the day lie just below surface with long antennae and tips of claws protruding. At night emerge to feed.

Atelecyclus rotundatus—Circular crab

Widespread but not often seen on very coarse sand, gravel or maerl sites deeper than 20 m. Often found with sea-cucumber *Neopentadactyla mixta*. Less than 4cm in diameter.

Goneplax rhomboides— Mud runner or square crab

Local but abundant where present living in burrows in fine mud at all depths. Run very fast with legs extended. Eyes on 'hinged' stalks. Colour variable, yellow to purple.

Liocarcinus puber—Velvet swimming crab

Widespread in rocky areas with a lot of water movement. Can swim fast sideways using flat hind legs as paddles. 'Rears up' in a threat posture when disturbed.

Liocarcinus depurator—Swimming crab

Common on fine sand, muddy sand and mud below about 10 m. Efficient swimmer using rear 'paddles', with characteristic purple spot on blade of paddle. 5–10cm diameter.

Carcinus maenas—Shore crab

Widespread and common on shores under stones, beneath seaweed and in rock pools. Subtidally to about 60 m. Especially common on mud in estuaries and sea loughs.

Cancer pagurus—Edible crab

Common in cracks and crevices in rocky areas with a lot of water movement. Also in sediment areas where they make and lie buried in a shallow depression or burrow under boulders.

Maja squinado—Spiny spider crab

A large southern crab, restricted to south and west around British Isles. Lives in sand abraded rock areas. Sometimes found in large mating mounds of more than 100 crabs.

Hyas araneus—Sea toad or great spider-crab

A northern crab. Widespread around British Isles at all depths on mud and muddy sand, also found on rock. Often cover themselves with weed or sponge. Eyes can retract. About 10cm long.

Inachus dorsettensis—Scorpion spider-crab

Very widespread but not common on all types of bottom from 5 m down in depth. Relatively heavy front claws carried bent under. Body up to 3cms, whole crab about 15cm.

Macropodia rostrata—Long-legged spider-crab

Very widespread, found mainly on muddy sand at all depths. A very 'leggy' small crab with relatively very large claws. No eye sockets. Often cover themselves with weed.

Crangon crangon—Common shrimp

Abundant subtidally in sandy bays and estuaries. Also found on lower shore. Nocturnal, buries itself during the day. Usually walks along bottom. Usually 8cm or less.

Palaemon serratus—Common prawn

Common in south, in rock pools, amongst algae and in shallow water. Also on sand amongst eel grass. Almost transparent. Can change spot patterns to match background.

Pandalus montagui—Northern prawn or 'pink shrimp'

Common on gravel sand and mud bottoms at all depths. Generally northern distribution. Characteristic red lines on back. Usually about 8.5cm long.

Balanus crenatus—A barnacle

The commonest subtidal barnacle. Attached to rock and shell in relatively sheltered water. Up to 2cm diameter. Catches food with legs extruded through 'trap-door' plates.

Alcyonium digitatum—Dead man's fingers

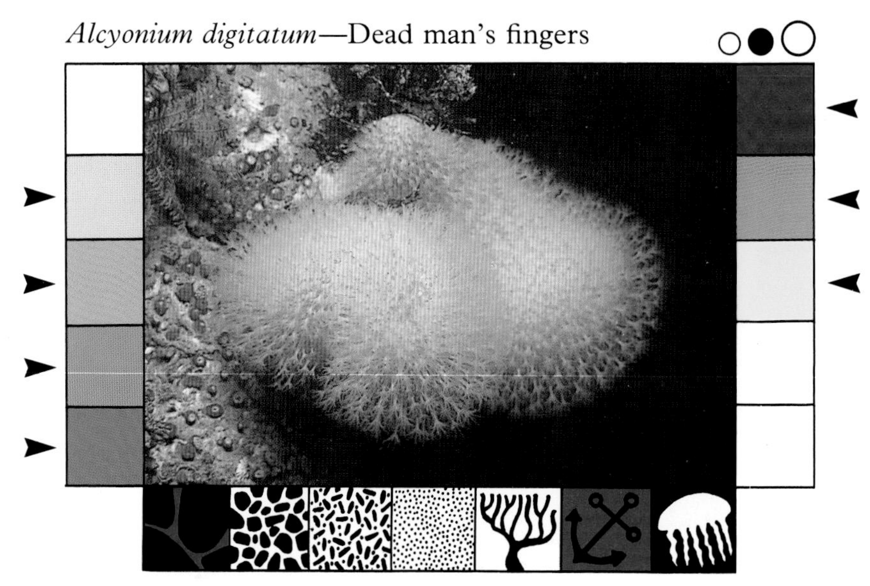

Colonise rock at all depths. Common in high water movement areas. Three colour forms—orange, beige and pure white. Feeding 'polyps' always white.

Alcyonium glomeratum—Red fingers

Always on rock in sheltered areas with little water movement. Live on vertical rock faces, under overhangs or in cracks and crevices. Western and southern.

Swiftia pallida—Northern sea fan

Slender white horny coral. Lives attached to deep rocks and boulders in sheltered areas. Only known in British Isles from the Scottish islands, west Scotland and SW Ireland.

Eunicella verrucosa—Sea fan

Orange pink or white horny corals live attached to deep rocks and boulders in south and west. Always orientate 'fan' across current. Slow growing—colony may be very old!

Virgularia mirabilis—Sea rush or Slender sea-pen

Widespread, and sometimes abundant in mud. Colony slightly 'dished' with feeding polyps on concave side orientated away from current. Can luminesce at night.

Pennatula phosphorea—Sea-pen

Widespread but local on deep mud sand or gravel on northern European coasts. Heavy red fleshy 'body' with white feeding polyps. Luminesces brightly when touched at night. Common in Scottish sea lochs.

Cerianthus lloydii

Widespread and sometimes abundant in mud, fine sand and muddy gravel at all depths. Make a mucus tube into which they can withdraw. Tentacles brown, white, green or banded.

Caryophyllia smithii—Devonshire cup-coral

True solitary coral with very variable colour. Live on rocks and boulders at all depths. In a few places with very stable sand they are found free on the bottom.

Corynactis viridis—Jewel anemone

Widespread on rock at all depths, sometimes abundant on west facing coasts. Forms 'sheets' of many individuals. Colour very variable. Characteristic knobs on tentacles.

Actinia equina—Beadlet anemone

The commonest anemone on shores and in rock pools on all coasts. Also found subtidally in shallow water. Attach to anything hard. Red, brown, orange or green colour.

Anemonia viridis—Snakelocks anemone

On rocks on shore and in shallow water. Tentacles not often retracted—even on stroking. Green, brown or grey colour, normally with purple tentacle tips.

Bolocera tuediae

Very large anemone (up to 30cm) attached to stones and shells in deeper water. White, pale pink or orange in colour. More common in north. Can 'pinch off' its tentacles.

Metridium senile—Plumose anemone

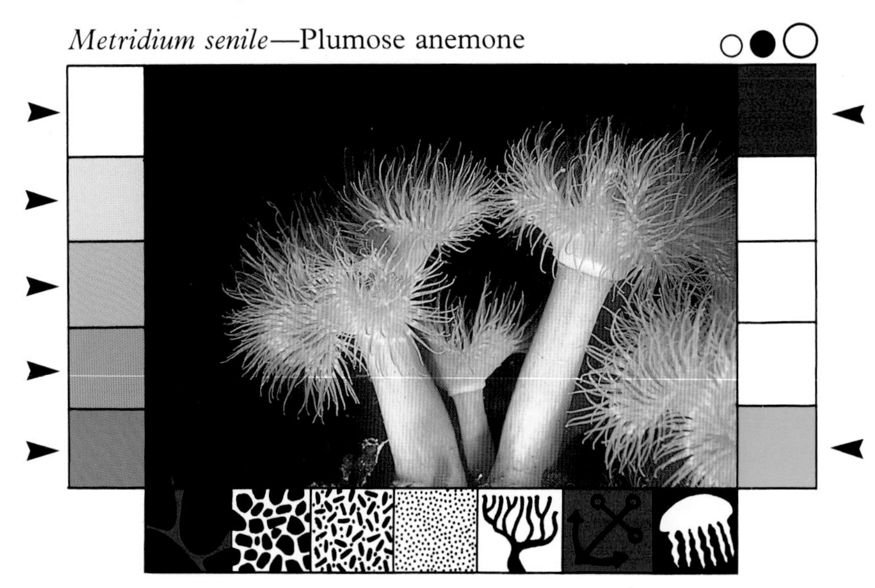

On all coasts on lower shore to deep water, often in very large numbers. Attached to rocks, wrecks, and piers in areas of strong water movement. White, yellow orange or brown.

Sagartia elegans

Very common at all depths on all coasts on rock or shell. Sometimes in large numbers. Extremely variable in colour. Throws out white threads when touched.

Actinothoe sphyrodeta

Widespread but local at all depths on smooth rock—not in cracks or holes. May be in large numbers. Disc can be orange yellow or white. Tentacles always white.

Sagartiogeton laceratus

Subtidal on all coasts, on stones or shell. Often attached to shell just below surface of mud or sand leaving only tentacles showing. Usually in small groups.

Calliactis parasitica—Parasitic anemone

Subtidal, mainly on southern coasts. Usually on shells inhabited by hermit crabs and sometimes on living whelk shells. Occasionally free on stones.

Adamsia maculata—Cloak anemone

Only found wrapped entirely around a species of hermit crab, *Pagurus prideauxi*. Small white anemone tentacles just under crab head to catch dropped food.

Tubularia indivisa

Lives on rock at all depths. In areas of strong current is often in large dense patches. Straight 'stems' usually rise from a mass of rock encrusting organisms.

Sertularia argentea—Sea fir

Widespread and common subtidally in areas with a lot of water movement. Fine stems with many branches. Can form very dense mats of bushy colonies. Brownish colour. Size up to 20cm.

Nemertesia antennina

Widespread as erect clusters of unbranched stems on stable subtidal rock at all depths. Can be abundant in strong current areas but not in wave exposed areas. Size up to 25cm.

Halecium halecinum

Widespread and common on subtidal rock or on shell on mud in sheltered areas. Can tolerate a lot of silt. Characteristic stiff stem with regularly spaced branches. Size up to 10cm.

Cyanea lamarckii

Found off all European Atlantic coasts. Four frilly mouth arms and a large number of long stinging tentacles. Size up to 30cm across. Very similar to larger brown *Cyanea capillata*—the Lion's mane.

Rhizostoma pulmo

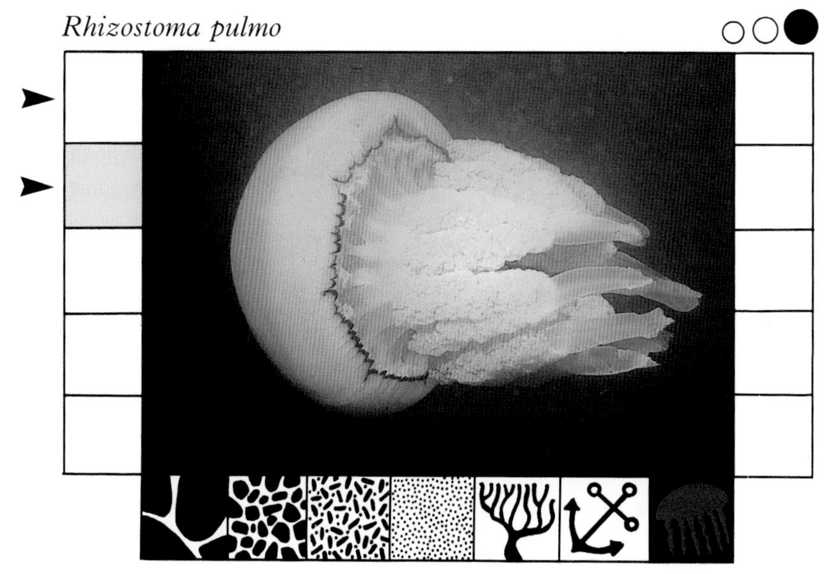

Found in all European seas. Almost opaque tall white domed bell with frilly purple edge, up to 1 m across. Eight characteristic partially fused mouth tentacles hanging down far below bell.

Pelagia noctiluca

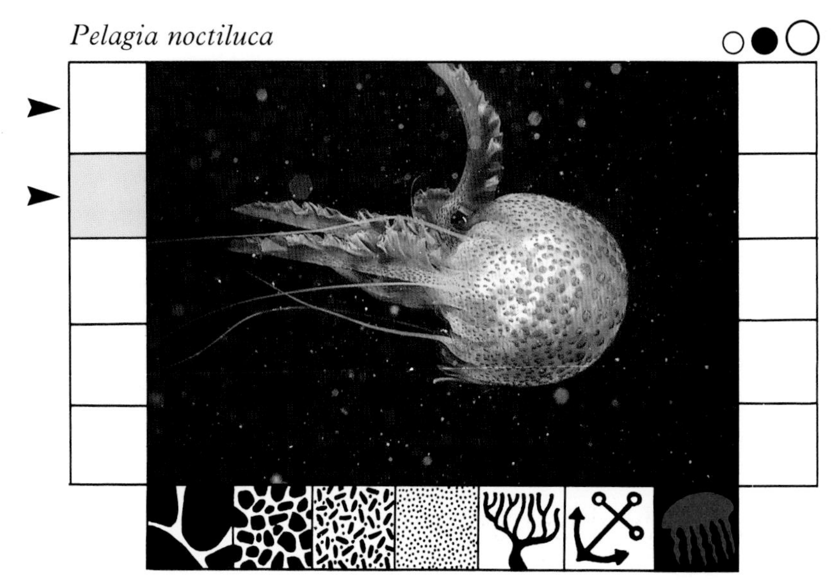

Widespread and sporadically abundant in late summer and autumn on all Atlantic coasts. Colour range includes pink, purple, brown and orange. Luminesces brightly at night when disturbed.

Chrysaora hysoscella—Compass jellyfish

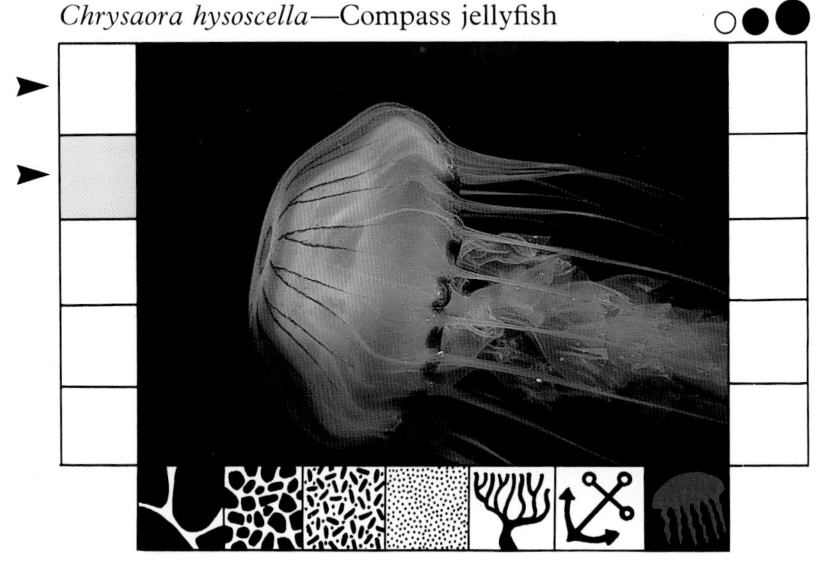

Found off all European Atlantic coasts. 'Compass' lines on dome. Four very long frilly mouth tentacles hang far below dome. Fringe of shorter tentacles around dome. Size up to 25cm across.

Apolemia uvaria—String jelly

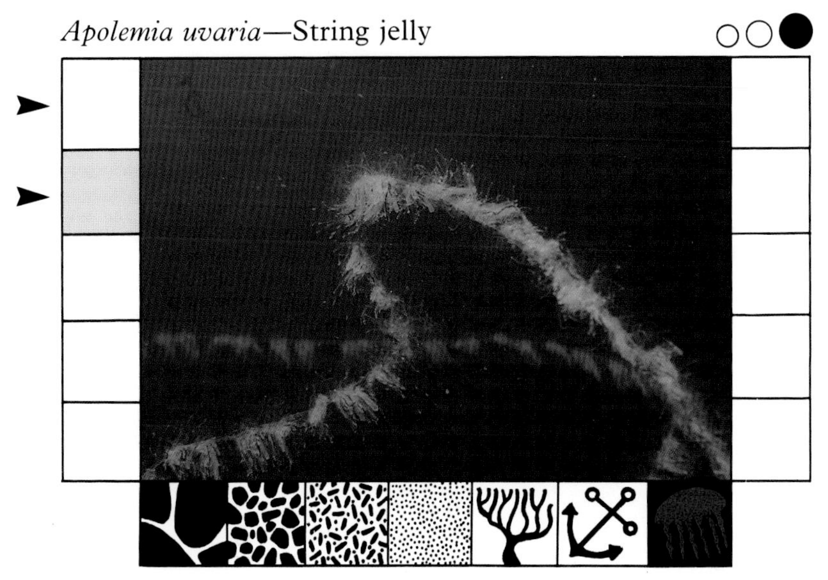

Long colony (up to 25m!) found drifting in all European seas. Small float at one end with trailing kinked string of transparent individuals. Often only spotted when string contracts.

Beroe sp—A comb jelly (*not* a Cnidarian)

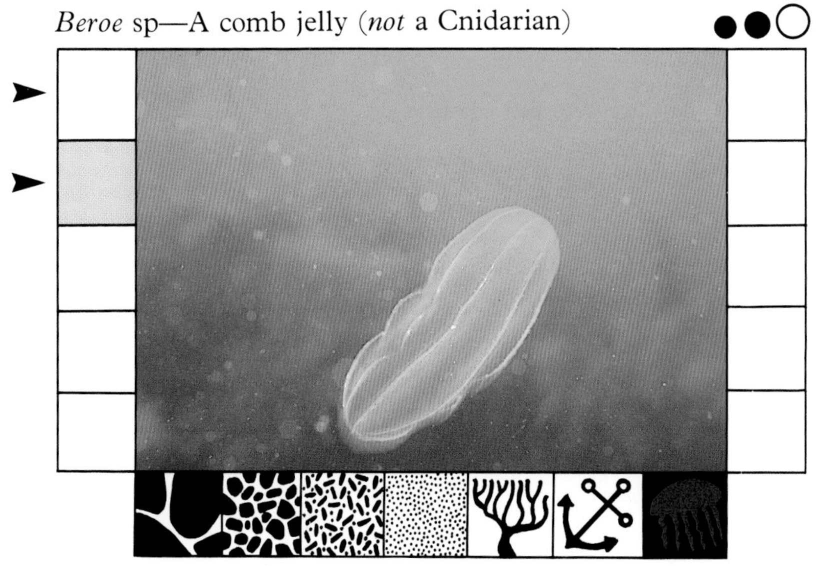

Widespread, sometimes in astronomical numbers in all European seas. Comb Jellies (Ctenophores) are *not* Cnidarians. They have 8 rows or combs of swimming plates. They do not have stinging cells.

Tonicella marmorea—A chiton or coat of mail shell

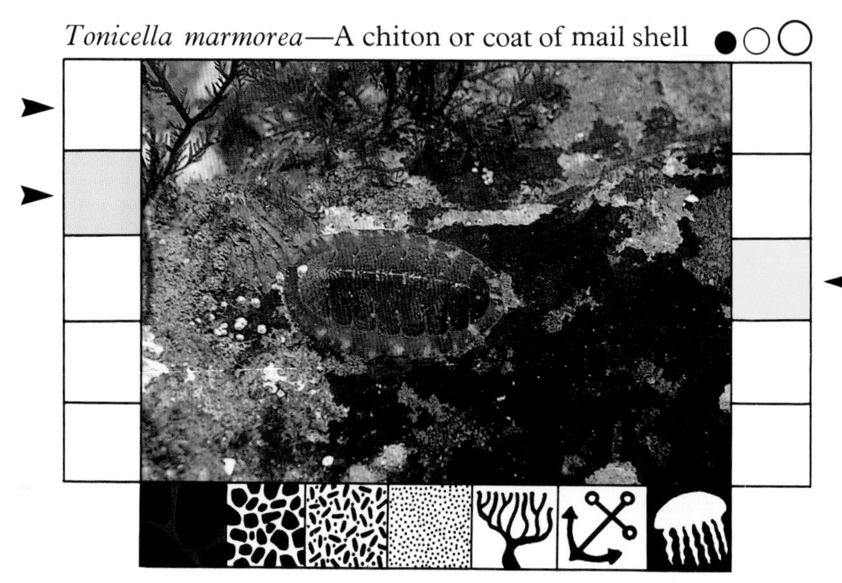

Widespread and common, on and under subtidal rock or old shell. Sometimes on shore. Shell made of eight interlocking plates. Roll up when taken off rock. Size up to 3cm.

Diodora graeca—Keyhole limpet

Widely distributed on rocks on lower shore and subtidally to about 20m. White breathing tube (siphon) pokes through 'keyhole'. Body tissue covers fringe of shell.

Patina pellucida—Blue-rayed limpet

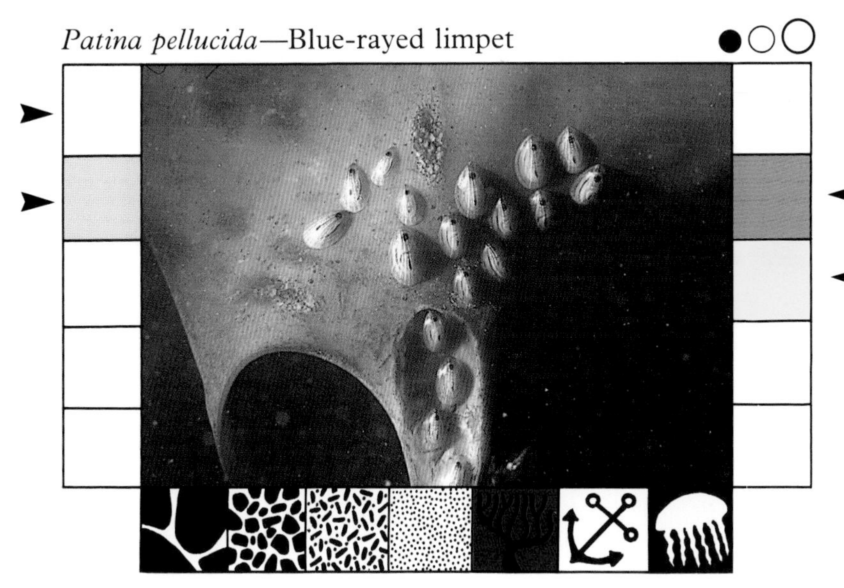

Widespread and common on kelp plants. Often in small groups at the base or near the junction of the 'stem' and 'leafy' part. Excavate individual small crypts into plant.

Calliostoma zizyphinum—Painted topshell

Widespread and common on lower shore and all depths below. Found in a wide range of habitat but most common on rock. A pure white form exists—in fast water?

Gibbula cineraria—Grey topshell

Widespread and common on lower shore and on kelp subtidally. Close red and white stripes produce a greyish effect. Under the shell, in the centre is a small hole. About 2cm high.

Gibbula magus—Large topshell

Widespread and common subtidally on mud sand and gravel in south. Also occurs further north on west facing coasts. Eight obvious whorls. Large hole in centre under shell. About 2cm high.

Turritella communis—Auger shell

Common, often in large numbers on fine sand and mud in deep water. In sea loughs may occur in fairly shallow water. Empty shells often used by young hermit crabs as temporary homes.

Aporrhais pespelecani—Pelican's foot shell

Widespread and common on fine sand and mud in moderately deep water. Usually covered with layer of mud and microscopic organisms. 'Foot' prevents sinking in mud.

Trivia monacha—European cowrie

Widespread and common on rock on lower shore and below. Feeds on small colonial sea squirts. Shiny shell with 3 dark spots completely covered by body when active. No spots—probably *Trivia arctica*.

Natica catena—Large necklace shell

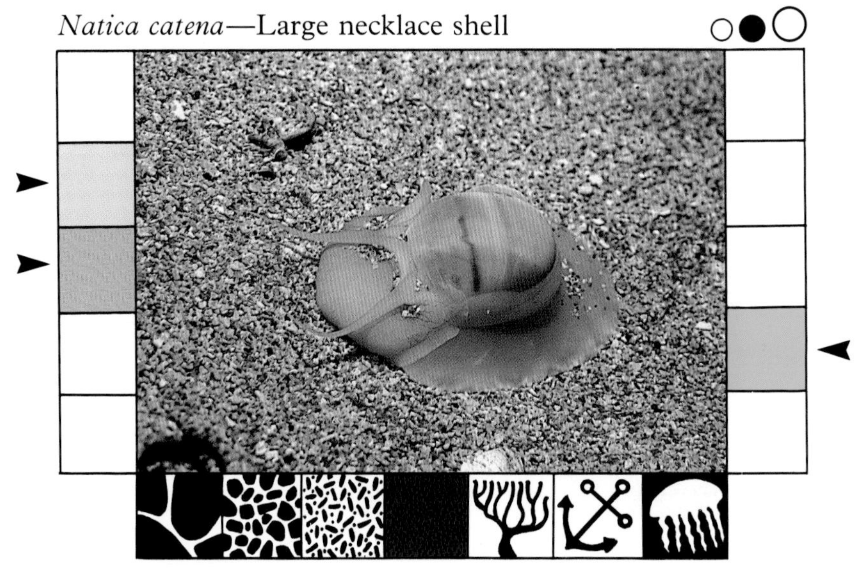

Widespread but local subtidally on sand. When undisturbed a large extension of the body spreads out across the sand and over the shell. Up to 5cm high.

Ocenebra erinacea—Sting winkle or Oyster drill

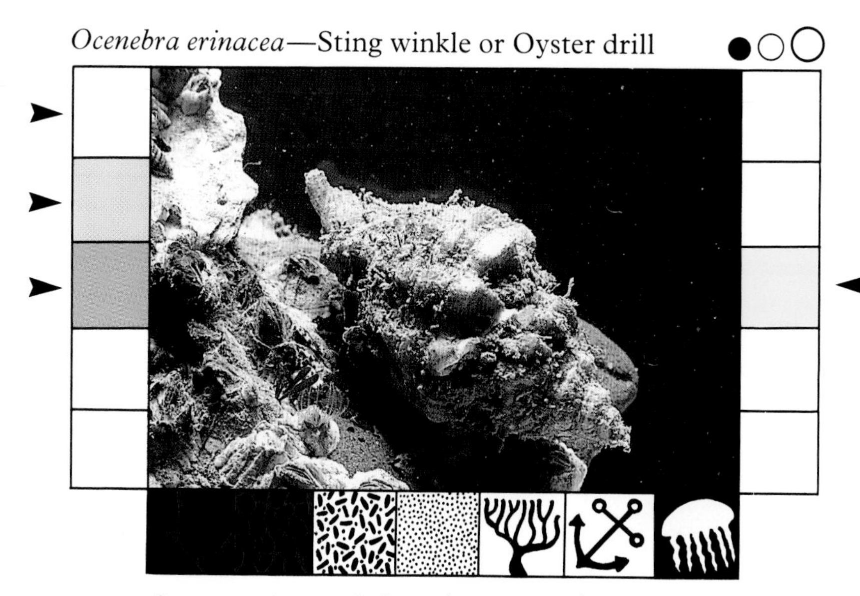

Common in south but also occurs in north. On rock on lower shore and shallow water subtidally. A major pest of Oyster beds. Lays yellow eggs in rock cracks. Size about 5cm.

Colus gracilis—Spindle shell

Mainly northern on deep current swept gravel or coarse sand. Elongated spire on spindle-shaped shell with many fine spiral ridges. Siphon emerges from tube in front of shell. Size about 5cm.

Buccinum undatum—Common whelk

Large (up to 16cm) common and widespread species on subtidal sediment. Rough ribbed shell. Long siphon tube is extended to sample water and detect food at distance. Smooth shell?—probably *Neptunea antiqua*.

Philine aperta

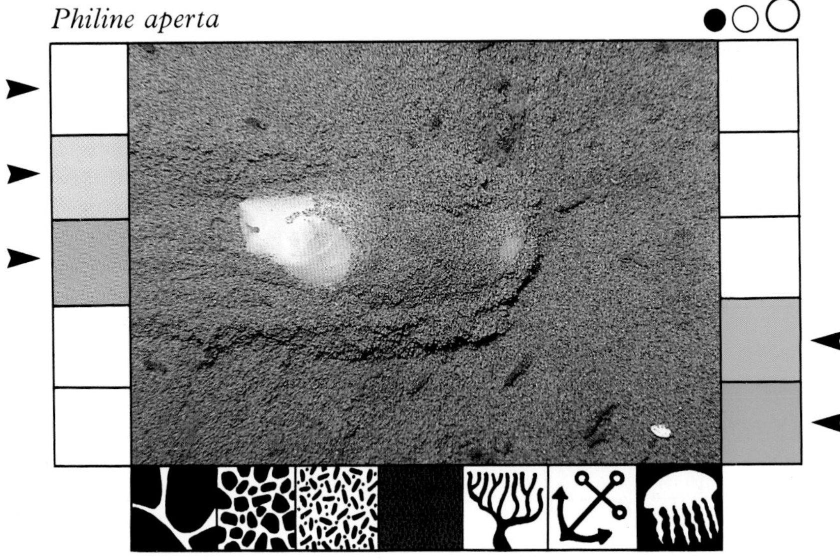

Extremely common and widespread in level fine sand and mud. Shell is internal, within the body. Burrows in surface of sand for small snails and worms. Often leaves 'trails' on surface.

Aplysia punctata—Sea hare

Widespread on rocky shores and in shallow water. Feed on seaweed. Seasonal, appear on shores in spring. Shell is internal. Colour black, brown or red. Do not swim. About 15cm long.

Elysia viridis—Green sea slug

Widespread on subtidal algae. Colour varies with food, green to red. Characteristic small shiny green red and blue spots on back. No shell. Small size (up to 2cm).

Pleurobranchus membranaceus

Common but sporadic on sheltered subtidal mud. A good swimmer (upside down!). Large gill on one side. Lays characteristic egg spirals. Skin can secrete sulphuric acid if attacked.

Dendronotus frondosus

Widespread as adults in fast water on *Tubularia indivisa* on which it feeds. Young feed on a variety of hydroids. Carry out unique threshing movement of body when disturbed. No shell.

Doto fragilis

Widespread subtidally on all coasts, feeding on hydroids like *Nemertesia antennina*. One of a group of small sea slugs, mostly less common, which feed on hydroids.

Acanthodoris pilosa

Widespread and common sea slug on shore and in shallow water. Feeds on Bryozoans like *Alcyonidium diaphanum*. Colour white, grey, brown or black. Gills at rear. Size 3–5cm.

Onchidoris bilamellata

Widespread on rock on shore and in shallow water. Feed on barnacles. Large number of gills in circle at rear. Skin can secrete acid when disturbed. Up to 4cm.

Archidoris pseudoargus—Sea lemon

Very large sea slug (up to 12cms). Common and widespread on shore and in shallow water. Feeds on sponges, mainly *Halichondria panicea*. Eight or nine bushy gills.

Coryphella lineata

Common and widespread subtidally in fast water areas. Feeds on *Tubularia* species. Characteristic white line runs down back from tentacles to tail. Size up to 5cm.

Facelina coronata

Common sea slug on shore and shallow water subtidally. Feeds on a variety of hydroids, but is found mainly on *Tubularia* species. Characteristic iridescent blue sheen.

Aeolidia papillosa—Grey sea slug

Large (up to 12cm), widespread, common species of rocky shores and all depths subtidally. Eats sea anemones, taking large bites from column. Colour grey or brown with white V on head.

Eubranchus tricolor

Widespread sea slug in rocky areas at all depths subtidally. Sometimes abundant in summer, feeding on colonies of the *Nemertesia* species. Size up to about 5cms.

Mytilus edulis—Common mussel

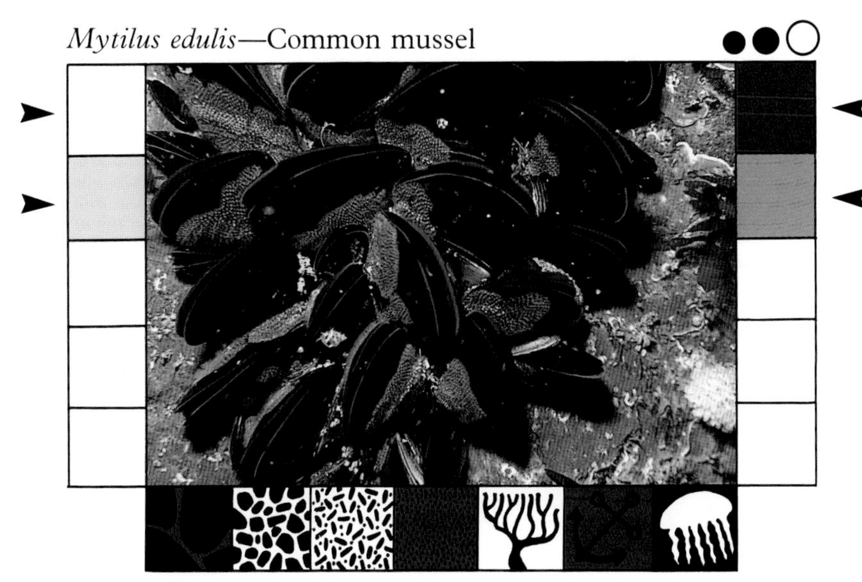

One of the commonest bivalve (two-shelled) molluscs of rocky coasts on shore and in shallow water. Also in vast beds on fine sand, in estuaries and sea loughs. Attach by 'byssus' threads.

Modiolus modiolus—Horse mussel

Large (up to 20cm). Widespread. Common in deeper water where it forms clumps in enormous beds. Many other animals live in and on clumps. Characteristic 'beak' near pointed end.

Pecten maximus—Great scallop, clam

Widespread and locally common on a wide range of sediment bottoms. Lies with curved (ash-tray) side down, often with gills facing current. Look for the smile! An active swimmer.

Aequipecten opercularis—Queen scallop

Widespread and sometimes abundant on subtidal mud. A very active swimmer. Equal sized 'ears' on shell. Both shells curved with about 20 ribs. Often has covering of sponge.

Chlamys varia—Variegated scallop

Widespread at all depths, attached to rock on shore or subtidally. Also free living or on *Modiolus* clumps. One 'ear' much bigger. Shells have about 25 ribs with spines. Size about 5cm.

Monia patelliformis—Saddle oyster

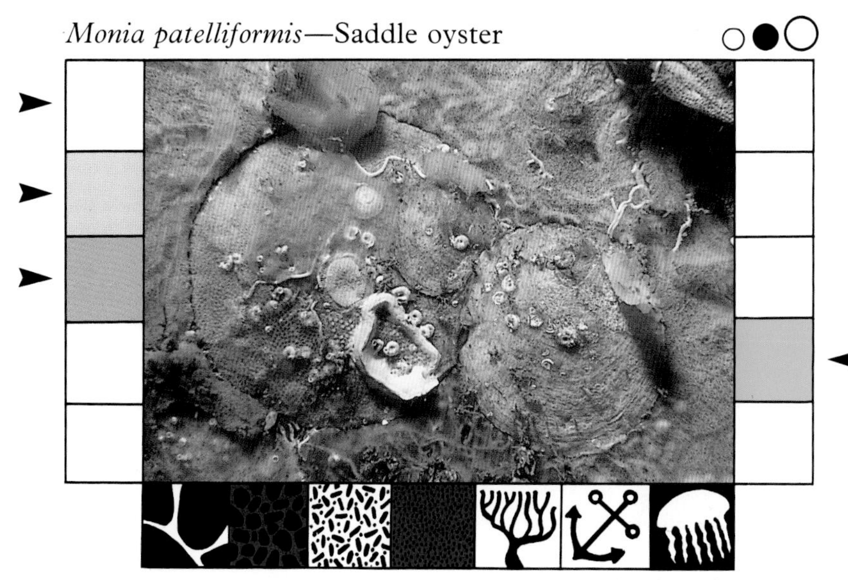

Widespread and common on all subtidal rock and shell, especially under boulders. Often have white chalky serpulid worm tubes attached. Similar, but larger, now rare—*Ostrea edulis*, the edible oyster.

Ensis siliqua—Razor shell 'keyhole'

Animals widespread in sand on lower shore and in shallow subtidal clean sand. Spotted on surface by 'Keyhole'—the feeding and breathing 'siphons' to surface through which water is pumped. Shells are well known 'razors'.

Circomphalus casina

Widespread and common subtidally in coarse sediment. Likes current swept areas. Sits in a characteristic position on its side, half out of the sand or gravel. Size about 4cm.

Sepiola atlantica—Little cuttle

Widespread and sporadically extremely numerous in shallow water over fine sand. Often 'sits' on bottom. When swimming uses two flat wings. Usually has a green 'eyebrow'. Size up to 5cm.

Eledone cirrhosa—Curled octopus

Widespread and common, particularly in north. Lives in holes and crevices in subtidal rock. Can change colour quickly and discharge 'ink' when disturbed. Single row of suckers. Size up to 50cm.

Crania anomala—A lamp-shell (*not* a mollusc)

Crania and *Terebratula* are *not* molluscs but Lamp-shells (Brachiopods). *Crania* is locally common on sheltered rock but not often recorded because it is difficult to spot. Size 1cm.

Terebratulina retusa—A lamp-shell (*not* a mollusc)

Widespread and sometimes numerous on deep current swept rock. Likes clean Atlantic water. Two convex 'shells' easily seen. Like all lamp-shells has a complex structure. Size 1cm.

Pachymatisma johnstonia—Elephant's ear sponge

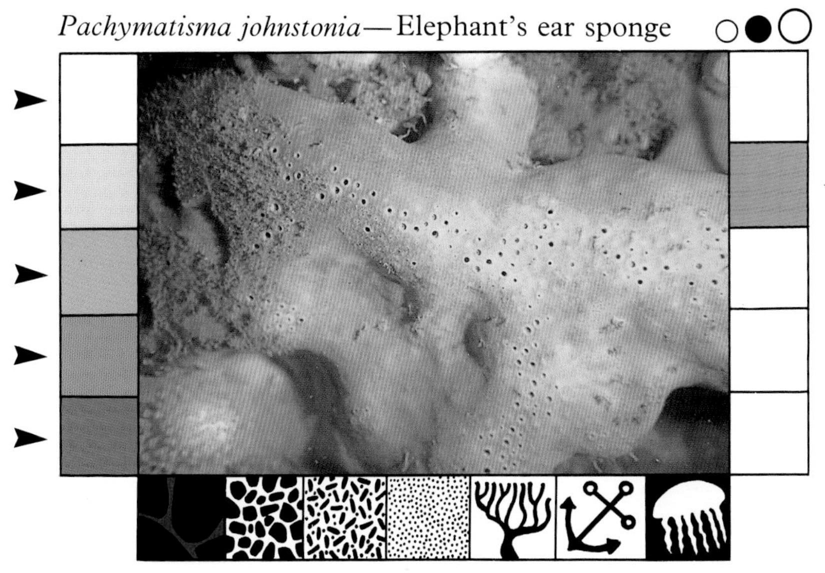

Widespread in south and west on vertical rock surfaces. Colonies are smooth and hard to the touch, are usually large and may be very large (up to 1m!).

Tethya aurantium

Widespread on rock, most common in deeper water. Rough looking ball with single 'hole' on top when undisturbed. Often covered by silt. Soft to the touch. Smells of decay?

Suberites carnosus

Widespread on stone and shell in mud. Smooth velvety ball with attachment 'stalk'. Single 'hole' on top. Contracts dramatically when disturbed.

Polymastia boletiforme

Widespread at all depths on upward facing rock or on the tops of boulders. Clearly defined massive colony with finger-like projections which can contract.

Cliona celata—Boring sponge

Widespread on rock. Large and sometimes enormous (more than 1m) in areas of strong water movement. Starts life as an inconspicuous borer of shell and limestone.

Axinella infundibuliformis

Widespread, on all coasts at all subtidal depths. Most common on rock in deep water. Characteristic shallow wine-glass shape with attachment 'stem'.

Axinella polypoides

Common on upward facing deep rock. Mainly southern, in 'clean' water conditions near the open Atlantic. Oval section branches may join up. Colour, yellow or orange.

Halichondria bowerbanki

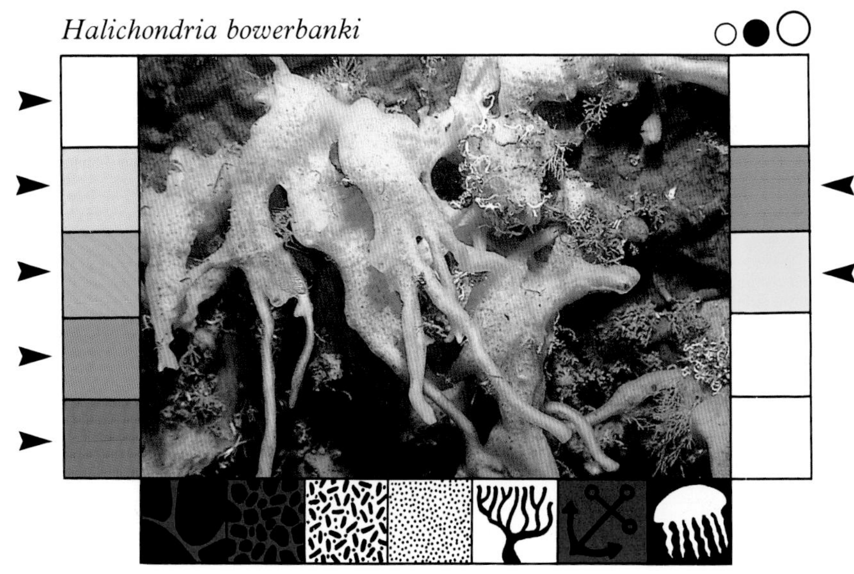

Very widespread and common subtidally. In sheets on rock and covering the surface of other animals. Never on shore. Colour grey, cream or buff, never green. Does not break when bent.

Halichondria panicea—Breadcrumb sponge

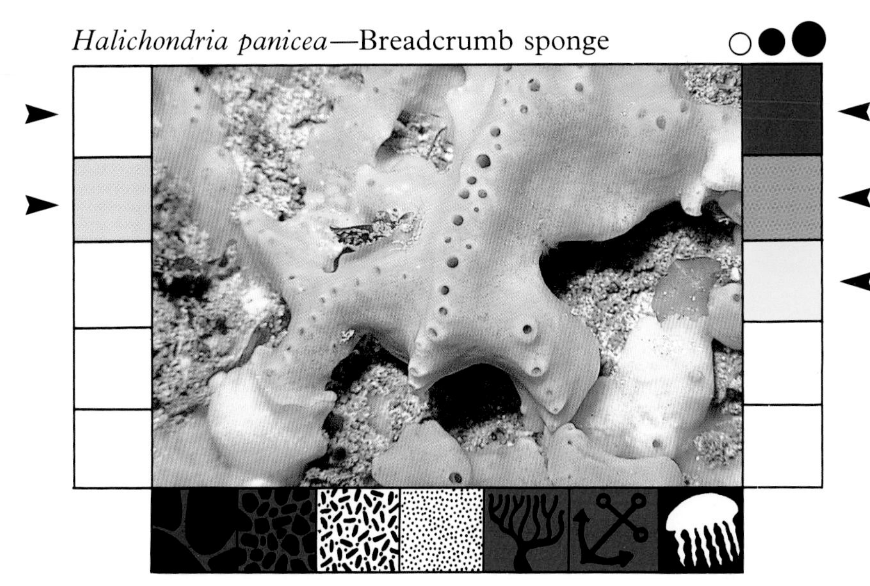

Shore and shallow water relative of *Halichondria bowerbanki*. Sometimes difficult to tell apart. Often in green sheets with obvious 'vents'. Breaks easily when bent. Distinctive smell!

Amphilectus fucorum

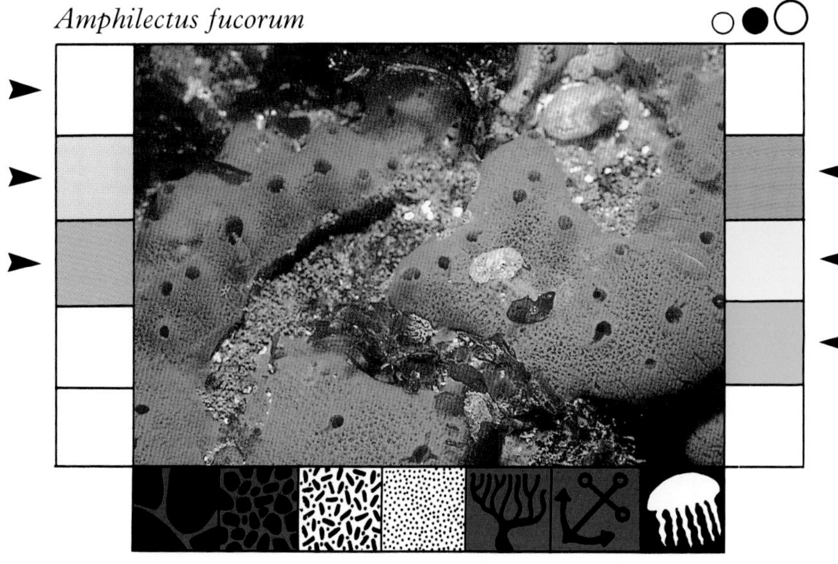

Small distinctive sponge. Widespread in moderate depths subtidally. Soft texture. Colour red or pale orange. Grows 'tassels' in calm conditions. Characteristic smell.

Myxilla incrustans

Very widespread on rock in clear water. Colony is a thick dense spreading cushion. Smooth to touch. Exudes a lot of slime when taken out of water.

Hemimycale columella

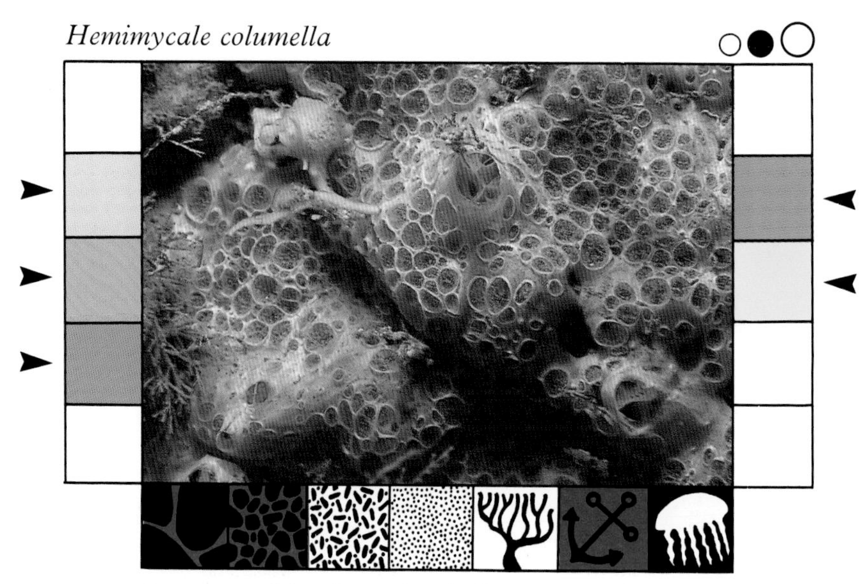

Occurs as thin sheets in shallow water. On clean rock and pebbles on all coasts, mainly in south. Very smooth surface. Characteristic pattern of circular pink depressions with pale rims.

Scypha ciliata—Purse sponge

Very widespread. Common on rock and weed on shore and in shallow water. Is sometimes found in deep water. Covered with hairs. Ring of long hairs at free end.

Clathrina coriacea

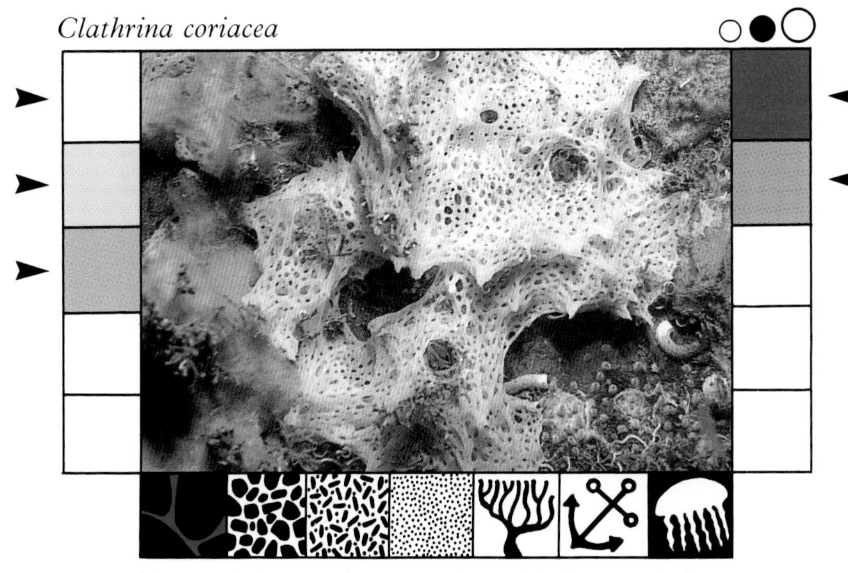

Very widespread on rock at all depths. Often found associated with the red sea squirt *Dendrodoa grossularia* on clean exposed rock. Colour, white, red, orange or yellow.

A Flatworm
Prosthecereus vittatus—Candy stripe flatworm

Widespread on all coasts amongst weed. Often on *Clavelina lepadiformis* (page 100) and on *Mytilus edulis* beds (page 79) on shore or in shallow water. Sometimes found on rocks in muddy areas. Size about 3cm.

A Nemertine worm
Tubulanus annulatus—Football jersey worm

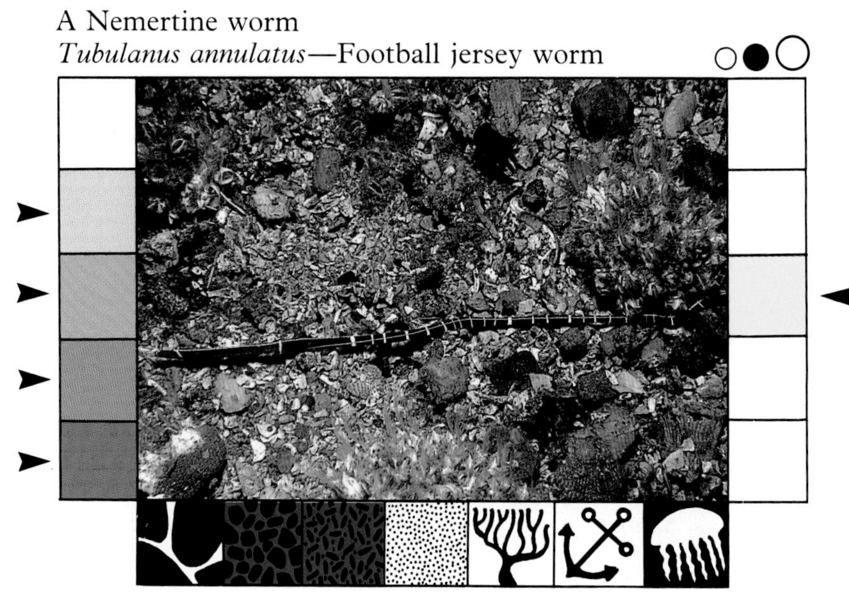

Widespread at all depths subtidally in sand and gravel. Also found under stones and in rock crevices. Usually about 12cm long but occasionally up to 1m!

Annelid worms

Arenicola marina—Lugworm (sand cast)

Widespread, common and sometimes abundant on sand shores and subtidally on sandy mud. Worm casts indicate presence of worm in U-shaped burrow in sand. Worms up to 20cm, casts usually about 10cm.

Aphrodite aculeata—Sea mouse

Widespread on subtidal mud, in deep water. Also on sheltered shallow mud. Matted fine brown hairs cover back. Coarse iridescent green blue and gold hairs on sides. Size about 10cm.

Eupolymnia nebulosa—Strawberry worm

Widespread and common under stones in muddy gravel in shallow water. Often only see dark blotched light tentacles writhing on surface. Body of worm is orange with white spots. Size up to 25cm.

Sabella penicillus—Peacock worm

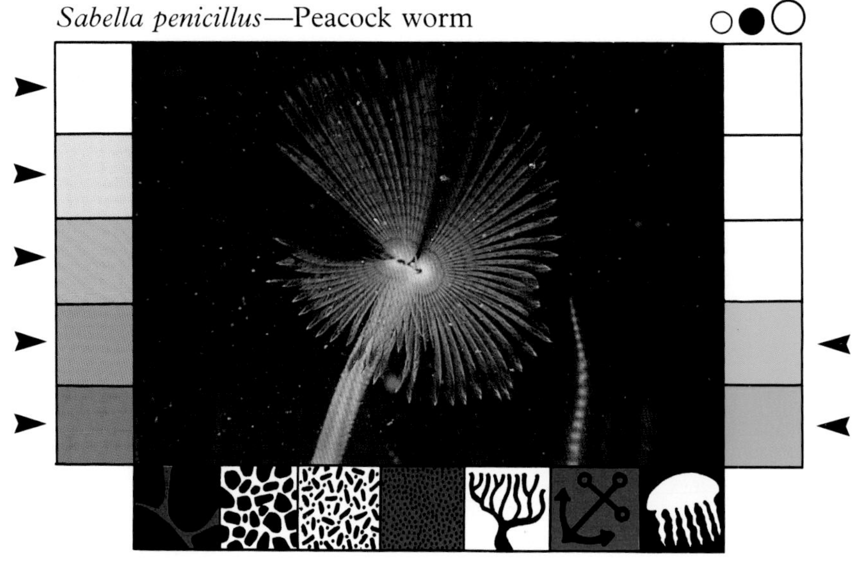

Common on lower shore and all depths subtidally. Mud and mucus tube is attached to stones or shell in mud or to deep rocks, mooring chains and wreck. Tentacles retract quickly. Size up to 25cm.

Bispira volutacornis

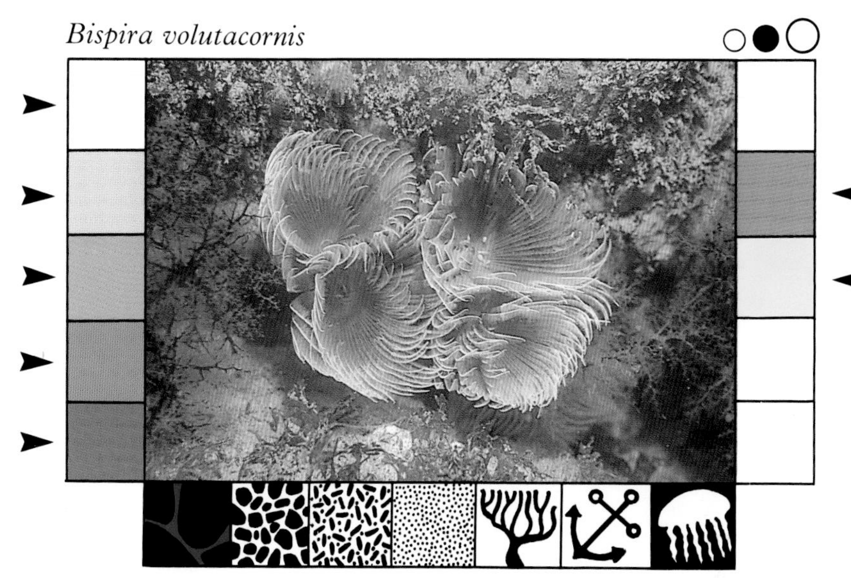

Common and widespread subtidally at all depths. Usually in holes and crevices in rock or attached to underside of cobbles. Thick crown of tentacles in two spirals about 3cm across.

Salmacina dysteri—Coral worm

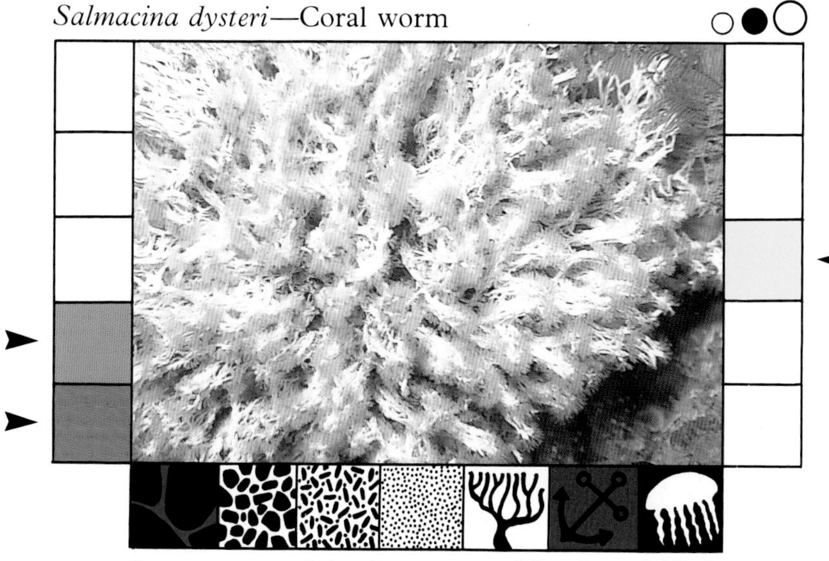

Scarce, on rock in deep water. Fine 'coral-like' brittle net of chalky tubes up to 20cm across. Worm tentacles emerge from tubes to feed. Shallow water, tubes more separated—probably *Filograna impexa*.

Lanice conchilega—Sand mason

Common and widespread in sand or among stones on lower shore and in shallow water. Normally only see characteristic worm tube made of similar sized large sand grains. Protrudes about 5cm.

Chaetopterus variopedatus—Parchment worm

Widespread in very coarse sand or gravel at all depths. Worm lies in a U-shaped, parchment-like tube which emerges 4cm above the surface at both ends. Tube often almost 50cm long.

Alcyonidium diaphanum

Common and widespread at all depths in a wide range of conditions. Colonies attach to rock, shell and kelp. Smooth surface with fine beard of tiny tentacles from individuals. On kelp—*A. hirsutum*.

Pentapora foliacea—Ross coral or Rose coral

At all depths on western facing, Atlantic coasts. More common in south. Very brittle. Initially attach to stone or shell in coarse sand. Grow to extensive colonies (up to 1m across).

Membranipora membranacea—Sea mat

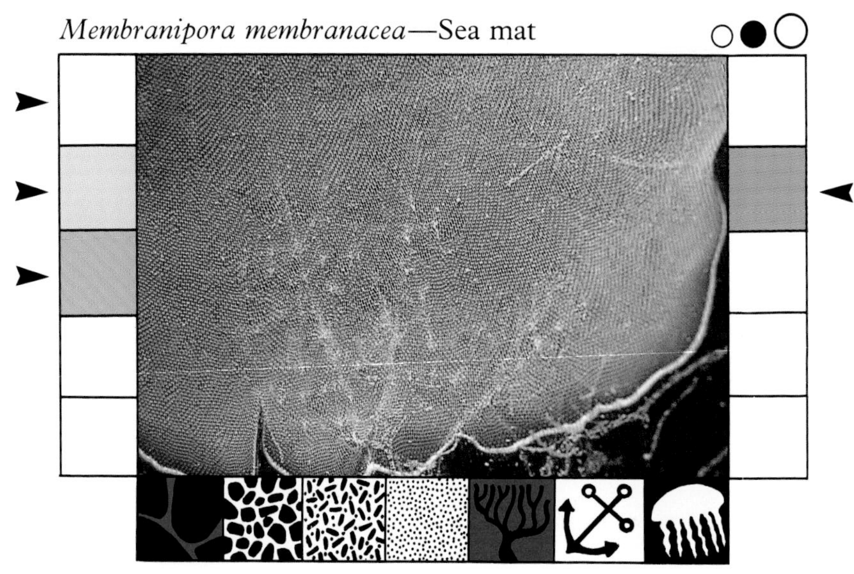

Very common and widespread as the lacy encrustation on the surface of kelp blades and stipes. Each segment is a complex individual member of the colony living in its own 'shoebox'.

Flustra foliacea—Hornwrack

Colonies widespread subtidally on rocks subjected to a lot of scour from sand in fast current areas. Sometimes in extensive beds. Smells strongly when ground between fingers, of lemon!

Cellaria fistulosa

Widespread and common in deeper water on a wide range of bottom types. Characteristic delicate two-way branching structure. Hard to the touch. Colonies up to 10cm high.

Bugula plumosa

Small colonies, widespread on lower shore and shallow water. Usually in small patches attached to rock, sometimes on overhangs. Characteristic spiral of branches. Size up to 5cm.

Clavelina lepadiformis—Light-bulb sea squirt

Live on rock at all depths on all coasts. May be abundant in moderately exposed shallow water sites. Each individual in colony is separate except at base.

Morchellium argus

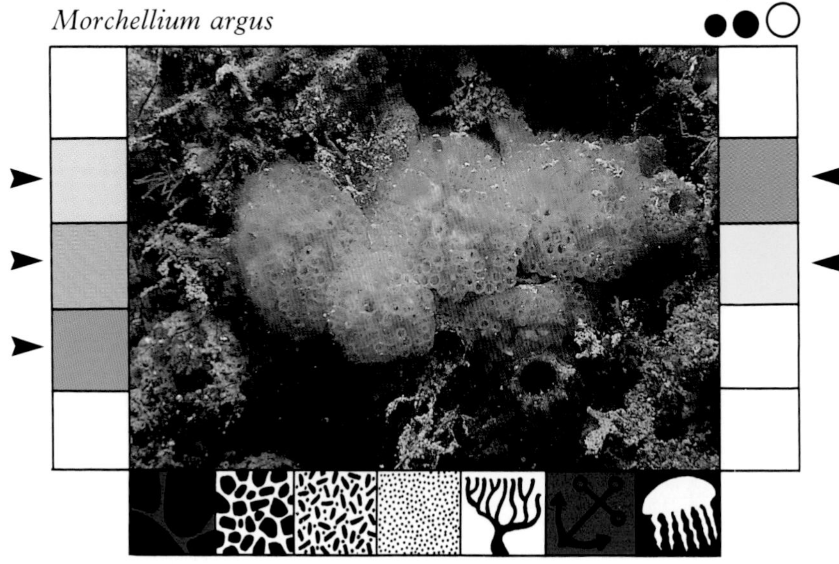

On rock in shallow strongly moving water on south and west coasts. Colonies have a distinct bulbous 'head' of transparent individuals and a red 'stalk' for attachment.

Aplidium punctum

Colonies at all depths on rock in fast moving water. Distinct 'head' and 'stalk'. Individuals in head each have an orange spot and extend into stalk as white lines.

Diazona violacea—Football sea-squirt

Large (10cm–40cm) ball shaped colonies on rock in deep clear water sites. White markings similar to *Clavelina lepadiformis*. Individuals joined over much of their length.

Corella parallelogramma

Widespread, at all depths on anything hard. Very transparent outer coat reveals characteristic geometric yellow or red criss-cross pattern on an internal sac.

Ascidiella aspersa

At all depths. In sheltered sites, often attached to shell or pebbles in mud. May be in large numbers. Can develop an attached coating of other animals and plants.

Ascidia mentula

Very widespread attached by one side to hard surfaces at all depths. Usually several individuals are grouped together. May be grey in absence of light in deep water.

Dendrodoa grossularia—Gooseberry sea squirt

Common and widespread in all situations. Particularly abundant as large sheets of individuals on exposed rock, in association with white sponge *Clathrina coriacea*.

Botryllus schlosseri—Star squirt ●●○

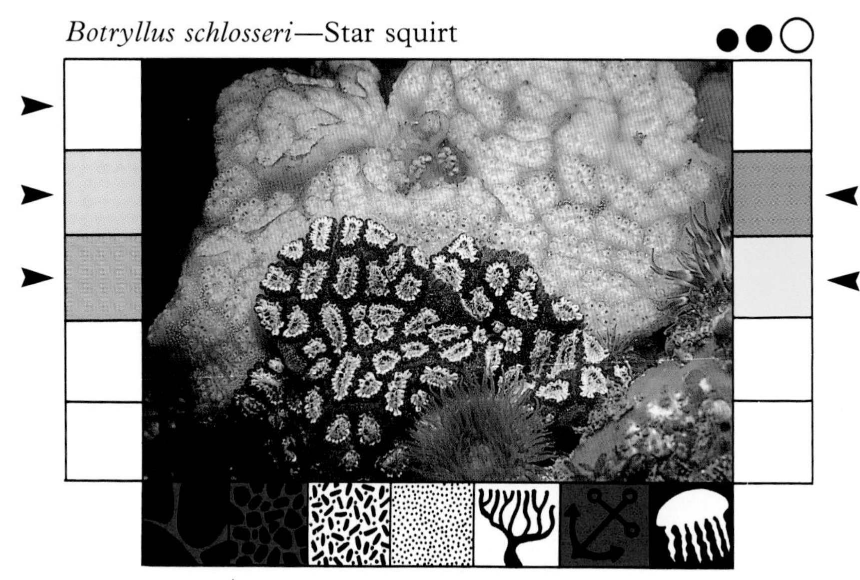

Colonies common and widespread on weed and rock on the lower shore and in shallow water. Occurs in a wide range of colours. Small groups of individuals in circles or 'stars' around common vents.

Botrylloides leachi ●●○

Colonies common and widespread on exposed rock surfaces on lower shore and in shallow water. Individuals in meandering lines. Colour usually orange, yellow or pink.

Brown algae

Laminaria hyperborea—Kelp or oarweed

Widespread and common. Forms classic 'kelp forest'. Rough stiff stalk or stipe encourages animal and plant cover. Two similar shallow water species with smooth stipes—*L. digitata* and in south *L. ochroleuca*.

Laminaria saccharina—Sugar kelp

Widespread. Common on rocks and boulders in mobile sediment. Attaches by 'holdfast' of round section branching strands. Single limp slippery blade with wavy edges. No mid-rib. Size 2–5m.

Saccorhiza polyschides

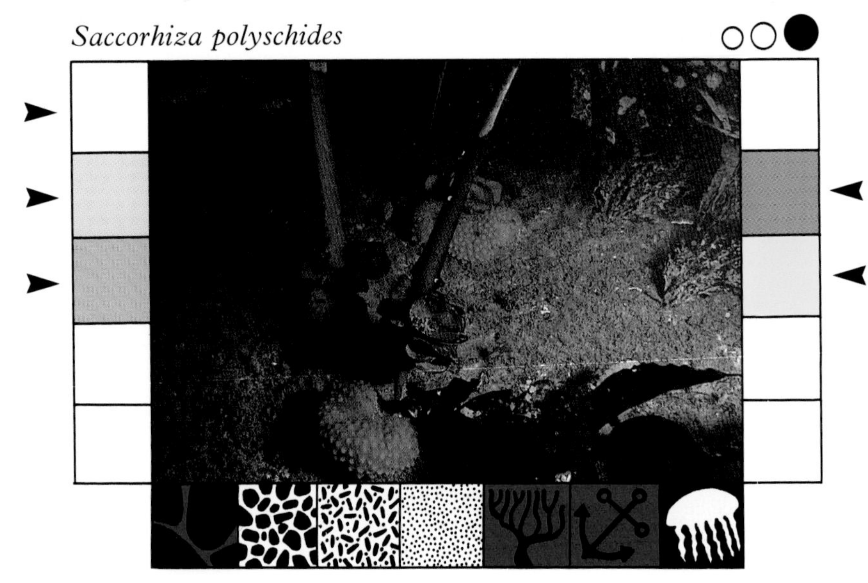

Widespread. Common spring to late autumn sheltered amongst *Laminaria hyperborea* on rock. Flattened stipe with 'frill' at base. Characteristic bulbous 'holdfast' for attachment. Size 2–5m.

Alaria esculenta

Widespread subtidally on rock in wave exposed areas. Blade like a *Laminaria saccharina* with light coloured midrib. Distinctive edible 'Keys' on stipe as in photograph. Size up to 5m.

Chorda filum—Sea whip

Widespread and common on small stones and shell in areas sheltered from wave action. Smooth, flexible and very slippery. Round section, unbranched and whip like. Size 1–3m.

Red algae
Drachiella spectabilis

Scarce but characteristic of Atlantic coasts. On bedrock in kelp forest and deeper. Plants flat, broad and almost transparent. In spring and summer have a unique blue iridescence. Size about 5cms.

Delesseria sanguinea

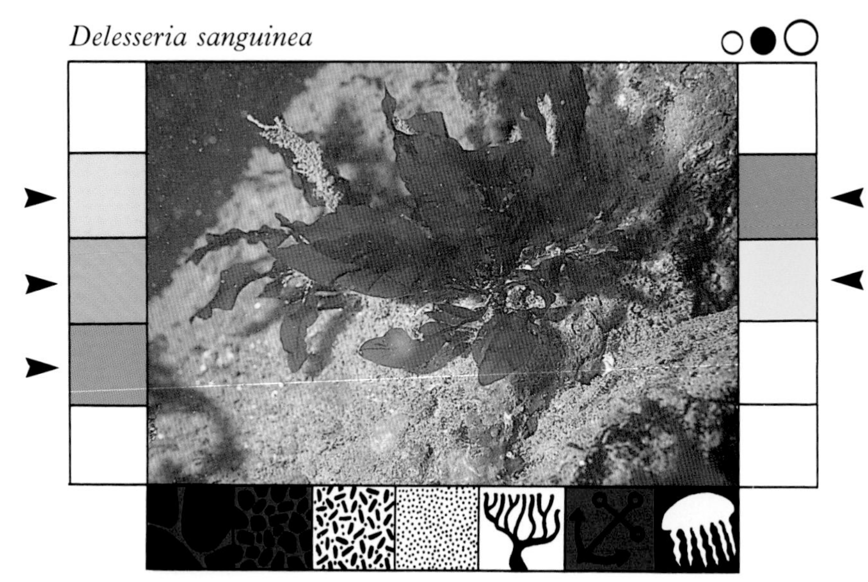

Widespread and common on bedrock in kelp forest and below. The bright red plants look like long copper beech leaves, complete with veins. They have a smooth unbroken outer edge. Size up to 30cms.

Dilsea carnosa

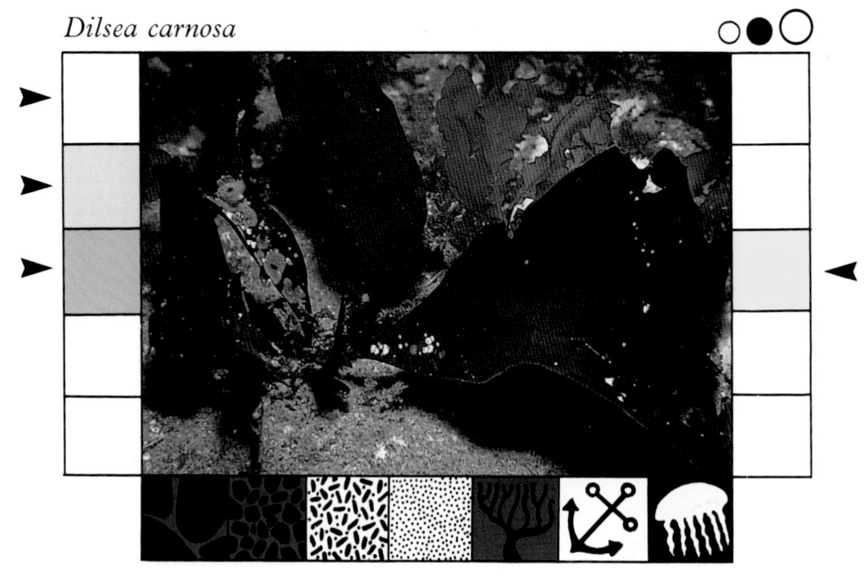

Widespread on rock in kelp forest. Thick leathery dark red leaves. Disc shaped holdfast and short stalks. Present all year round. Size up to 50cms.

Kallymenia reniformis

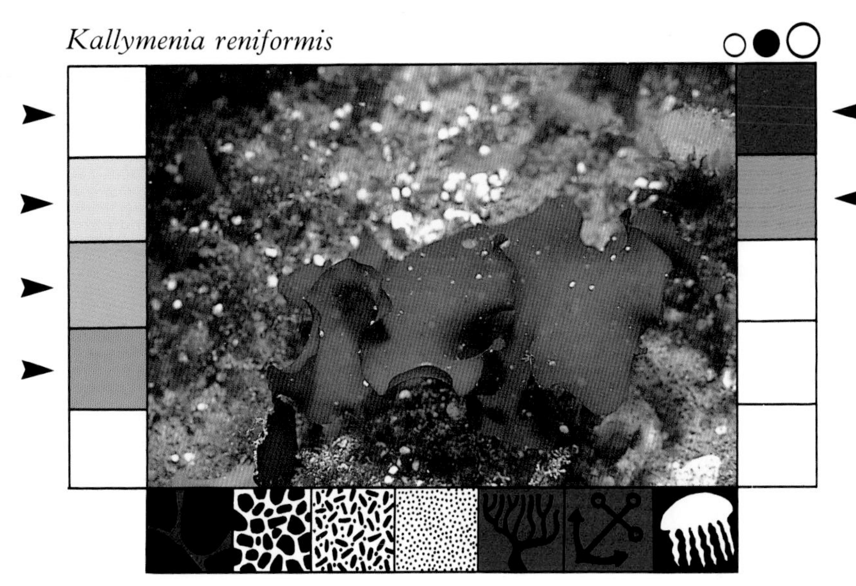

Widespread on rock in kelp forest. Thin liver coloured leaves on a short stalk. Shape of leaf very variable, usually roughly oval. Found throughout year. Size about 20cm.

Calliblepharis ciliata

Widely distributed subtidally. Common on shallow rock, and maerl. Occasionally in dense beds on rock below kelp forest. Thick brittle leaves have a fringe of small 'leaflets'. Size about 20cm.

Polyides rotundus

Widespread, with disc holdfast attached to rock or pebbles covered by sand. Slender cylindrical branching plant emerges through sand to produce a 'bushy' effect. Size up to 20cm.

Sphaerococcus coronopifolius

Widespread on south and west coasts. Large (up to 30cm) conspicuous plant. Very similar smaller and more common species—*Plocamium cartilagineum* with distinctive tips to branches like teeth of a comb.

Brongniartella byssoides

Widespread in summer intertidally, on rock in kelp forest and on mobile pebbles. Thick tufts of long fine stems with many alternating finely haired short side branches. Size up to 20cm.

Griffithsia flosculosa

Widespread in summer on kelp stipes and on rock in the kelp forest. Rigid tufts of long fine hair-like branching stems. Reproductive organs in circles of tiny branches on stems. Size up to 20cm.

Corallina officinalis—Coral weed

Widespread and extremely common on the lower shore and just below. Hard and brittle due to chalk skeleton of interconnecting beads. Branches always in opposing pairs. Size up to 10cm.

Maerl—Several difficult to identify plants

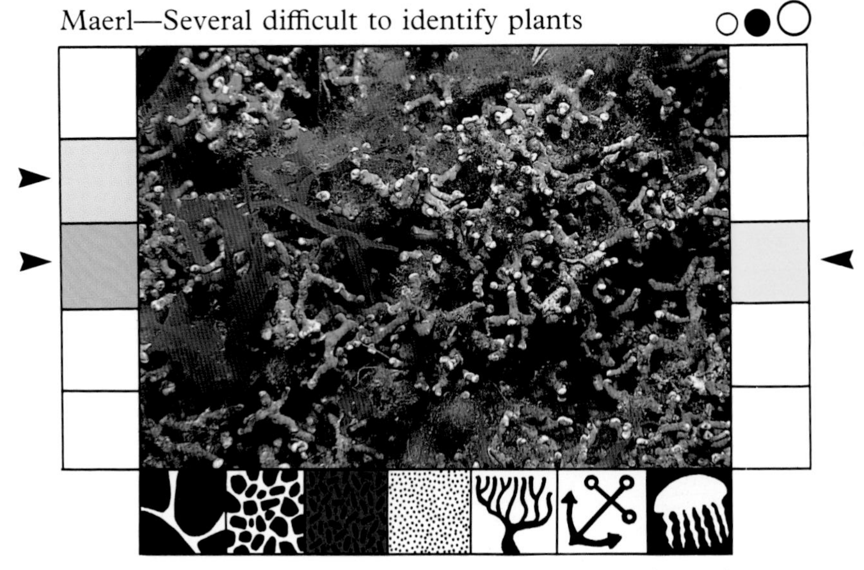

Widespread but most common in south and west facing coasts down to 25m. Extensive beds in wave protected current swept areas. Hard knobby particles of chalk with outer thin living layer.

Crustose corallines

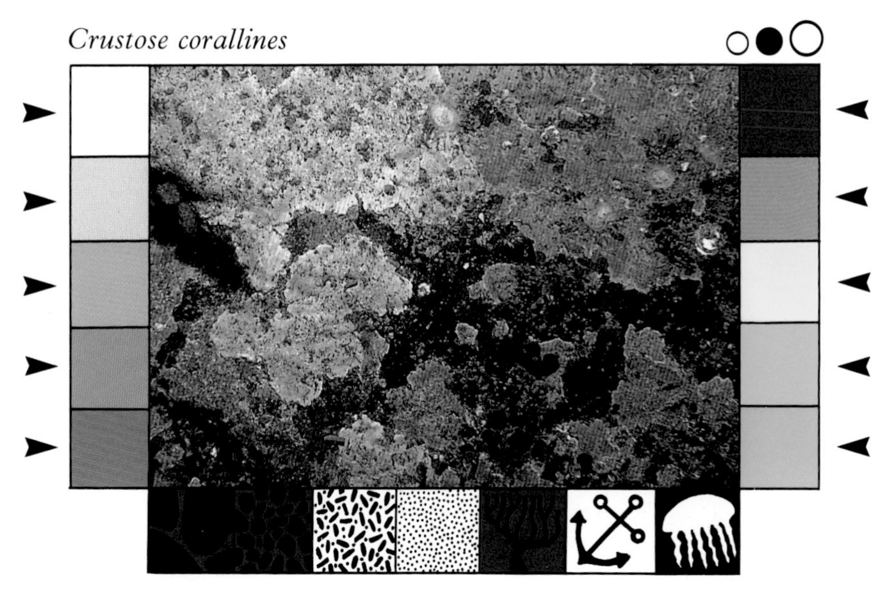

Occur on every hard material that light reaches. A large and very difficult to identify group of pink or red crusts. Vary in size from tiny patches to vast sheets on heavily grazed bedrock.

Green algae

Enteromorpha and *Ulva* species.—Sea lettuce

This and other very similar, difficult to identify species are widespread in shallow water in a wide range of conditions. All are either green tubes, or thin flat green sheets. Size up to 50cm.

Bryopsis plumosa

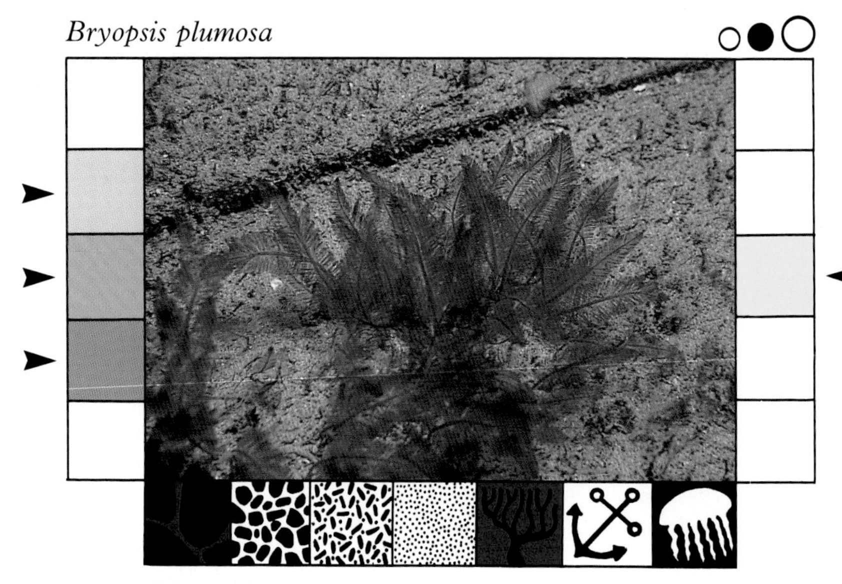

Widespread in summer, on rock in kelp forest, particularly on vertical faces. Also often found on deep rock. Flat stiff feather-like appearance with fine side branches in same plane and naked base.

Flowering plant

Zostera marina—Eel grass *not an alga*

Widespread, sometimes abundant in beds on fine sand in shallow water. One of a few species of marine plants with roots, leaves and true flowers—inconspicuous in leaf bases. Size up to 1m long.

Dr. David G. Erwin:
David Erwin is in charge of the Botany and Zoology department of the Ulster Museum in Belfast.

Bernard E. Picton:
Bernard Picton is employed as a curator of marine invertebrates at the Ulster Museum in Belfast.

ACKNOWLEDGEMENTS

The authors would like to thank Bob Earll and David George of the Marine Conservation Society for all their help and encouragement throughout the project; Jon Moore and Jim Adams for their help in selection of species and Christine Howson for providing several photographs.

We also wish to take this opportunity to dedicate this book to all those enthusiasts who freely and tirelessly gave of their time and effort in collecting records for the original "Species Recording Scheme".

FOR MORE INFORMATION READ:

A. Classification

1. Marine Life—An illustrated encyclopedia of invertebrates in the sea; by David and Jennifer George.
   Hardback: Lionel Leventhal Ltd., 1979.

2. Classification.
   Paperback: British Museum of Natural History, 1983.

3. Classification of the animal Kingdom—an illustrated guide.
   Paperback: Hodder and Stoughton Ltd., and Readers Digest Ltd., 1979.

B. General

1. Sea life of the British Isles; by biologists from the Marine Conservation Society. Paperback: Immel, 1988.

2. Basic marine biology; by Anthony Fincham.
   Paperback: British Museum (N.H.) and Cambridge University Press, 1984.

3. An introduction to coastal ecology; by Pat Boaden and Ray Seed. Paperback: Blackie, 1985.

C. General Identification Guides

1. Field Guide to the Water Life of Britain.
   Hardback: Readers Digest Nature Lovers Library, 1984.

2. The Country Life Guide to The Seashore and Shallow Seas of Britain and Europe; by Andrew Cambell.
   Paperback: Country Life.

D. Specialised Identification Works

There is a large and confusing array of specialist works available. To the general reader wishing to go further we recommend:

1. The Synopses of the British Fauna (New Series).
   Paperback: Current list from Linnean Society, Burlington House, Piccadilly, London, W1V 0LQ.

2. The AIDGAP guides.
   Paperback: Current list from AIDGAP, The Leonard Wills Field Centre, Nettlecombe Court, Williton, Taunton, Somerset, TA4 4HT.

These excellent series are still expanding and already cover many of the marine groups commonly found around European coasts. They are written by the authorities within each group yet retain a level of complexity well within the capacity of the enthusiast.

Attention is also drawn to the Miniprint sets of photographs and information available from the Marine Conservation Society. Sets now available include Sponges, Anemones, Nudibranchs (sea slugs), Echinoderms, Sea squirts and Algae.

INDEX